CHEMISTRY
IN ACTION

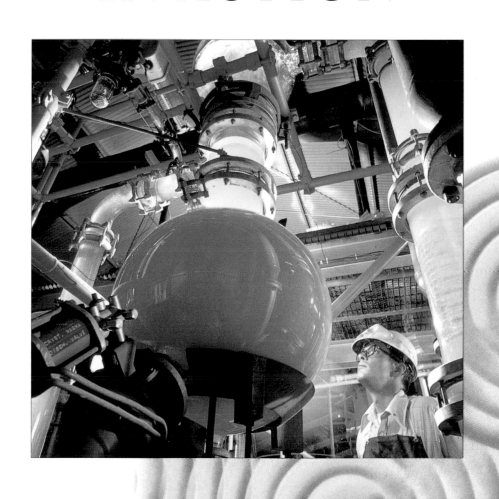

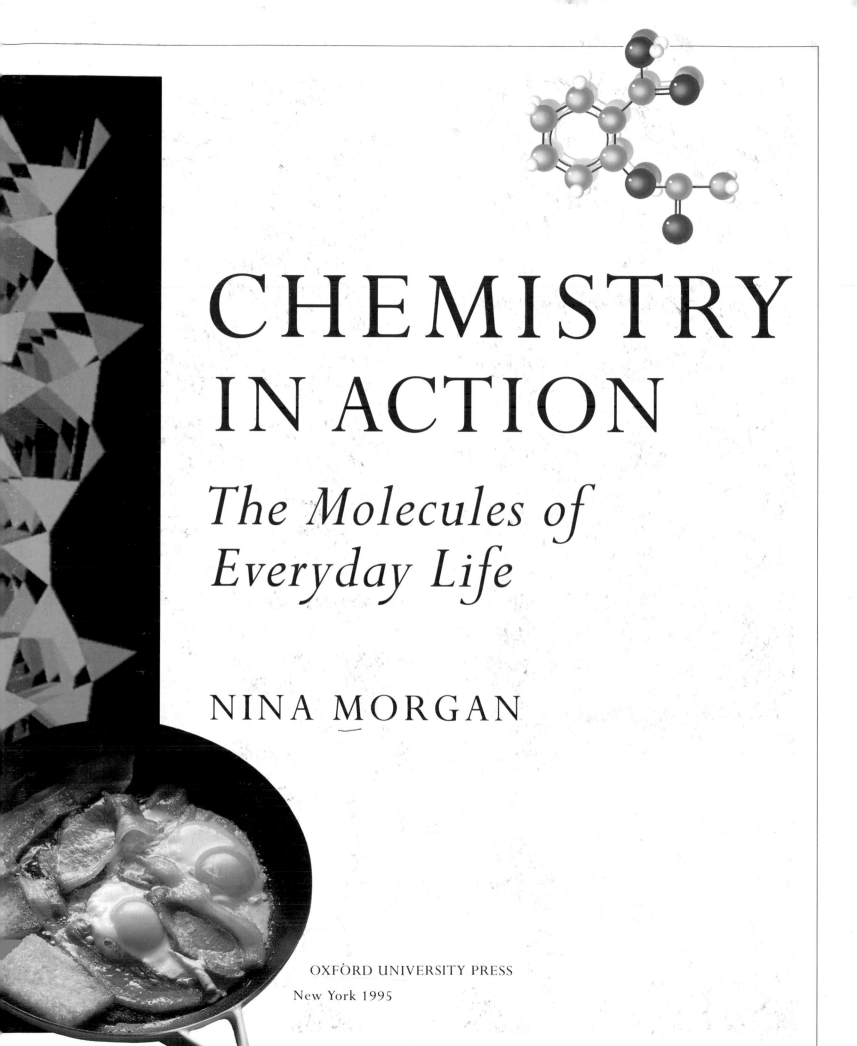

CHEMISTRY IN ACTION

The Molecules of Everyday Life

NINA MORGAN

OXFORD UNIVERSITY PRESS

New York 1995

CONTENTS

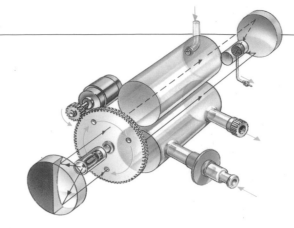

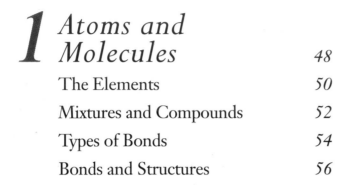

Project editor	Peter Furtado
Senior editor	John Clark
Editor	Lauren Bourque
Editorial assistant	Marian Dreier
Art editor	Ayala Kingsley
Visualization and artwork	Ted McCausland/ Siena Artworks
Senior designer	Martin Anderson
Designer	Roger Hutchins
Picture manager	Jo Rapley
Picture research	Jan Croot
Production	Clive Sparling

Planned and produced by
Andromeda Oxford Ltd
9-15 The Vineyard
Abingdon
Oxfordshire OX14 3PX

© copyright Andromeda Oxford Ltd 1995

Text pages 16-47
© copyright Helicon Ltd,
adapted by Andromeda Oxford Ltd

Published in the United States of America by
Oxford University Press, Inc.,
200 Madison Avenue
New York, NY10016

Oxford is a registered trademark of Oxford University Press

Library of Congress Cataloging-in-Publication Data

Morgan, Nina
   Chemistry in action: molecules in everyday life / by Nina Morgan
        p.   cm. -- (The new encyclopedia of science)
   Includes bibliographical reference (p. 153) and index
   ISBN 0-19-521086-7   : $35.00
   1. Chemistry   2.   3.   I. Title   II. Series.
QD33.M79   1995
540--dc20   94-30086 CIP

Printing (last digit):9 8   7 6 5 4 3 2 1

Printed in Spain by Graficromo SA, Cordoba

INTRODUCTION

THE WORD "CHEMICAL" is loaded with negative meaning for more and more people in industrial societies, who feel desperately worried about the impact of synthetic chemicals on the natural world. Yet nature itself is the greatest chemist. We breathe O_2, drink H_2O and sprinkle NaCl on our food. Each of these – air, water and salt – is a chemical compound, made from naturally-occurring chemical elements. There are about 90 of these chemical building blocks that can be found in nature, in addition to a dozen or so that can be synthesized in a laboratory. Only about 30, though, are widely found. From these 30 building blocks, nature can assemble millions of different substances. Chemists have studied how these substances are built and how their structures relate to their interactions with other substances. They have also learned to build new substances of ever-increasing complexity to assist us in feeding ourselves, building our shelters, keeping ourselves healthy, permitting safe travel, and many other tasks. Despite the genuine threats represented by dangerous or toxic waste products from chemical processes, the overall impact of chemistry on our world is overwhelmingly positive.

Elements combine into compounds when their atoms form chemical bonds with those of other elements, and compounds react by breaking some bonds and forming others. Chemical bonds themselves involve the exchange or sharing of electrons between atoms. Chemistry is essentially about the behavior of the atoms and electrons of the elements, and how they interact.

More than eight million compounds are known to chemists. Only one million of these were known 60 years ago; the rest have been discovered or invented since then. Most compounds are organic – that is, they contain carbon. Chemists once believed that carbon compounds could only be made by living things, such as the cells in an animal's body, but most organic compounds today are built in the laboratory. Organic compounds include hydrocarbons (compounds of hydrogen and carbon, such as fossil fuels); carbohydrates (such as sugars, starch and cellulose); polymers and plastics.

Inorganic chemicals generally have no carbon in them, although they include a few carbon compounds, such as the oxides of carbon – carbon monoxide and dioxide – and carbonate and bicarbonate salts. Acids, alkalis and salts are important inorganic compounds, as are all ores and minerals (except fossil fuels).

Chemists need to be able to analyze substances to find out what they contain; to devise processes to create required substances cheaply and safely out of the available raw materials; and to predict the properties of a new compound before they go to the expense of creating it. Chemistry therefore combines the theoretical and the practical, the laboratory and the factory, in a unique way. As a chemist once remarked: "Chemists are the only scientists who make things; the others merely theorize, analyze or dissect."

Chemistry combined with other sciences extends the possibilities for applying the chemist's knowledge. Biochemistry is increasingly important in medicine, and contributes to the understanding of life itself. Physical chemistry – the study of the physical, rather than chemical, properties of compounds – includes electrochemistry, thermochemistry and reaction kinetics (the factors that affect the speed of chemical reactions). Industrial chemistry, which includes chemical engineering, is the most commercially important branch of chemistry. It gives rise to useful products such as ceramics, cosmetics, dyes, explosives, fertilizers, medicines, paints, silicon chips and plastics.

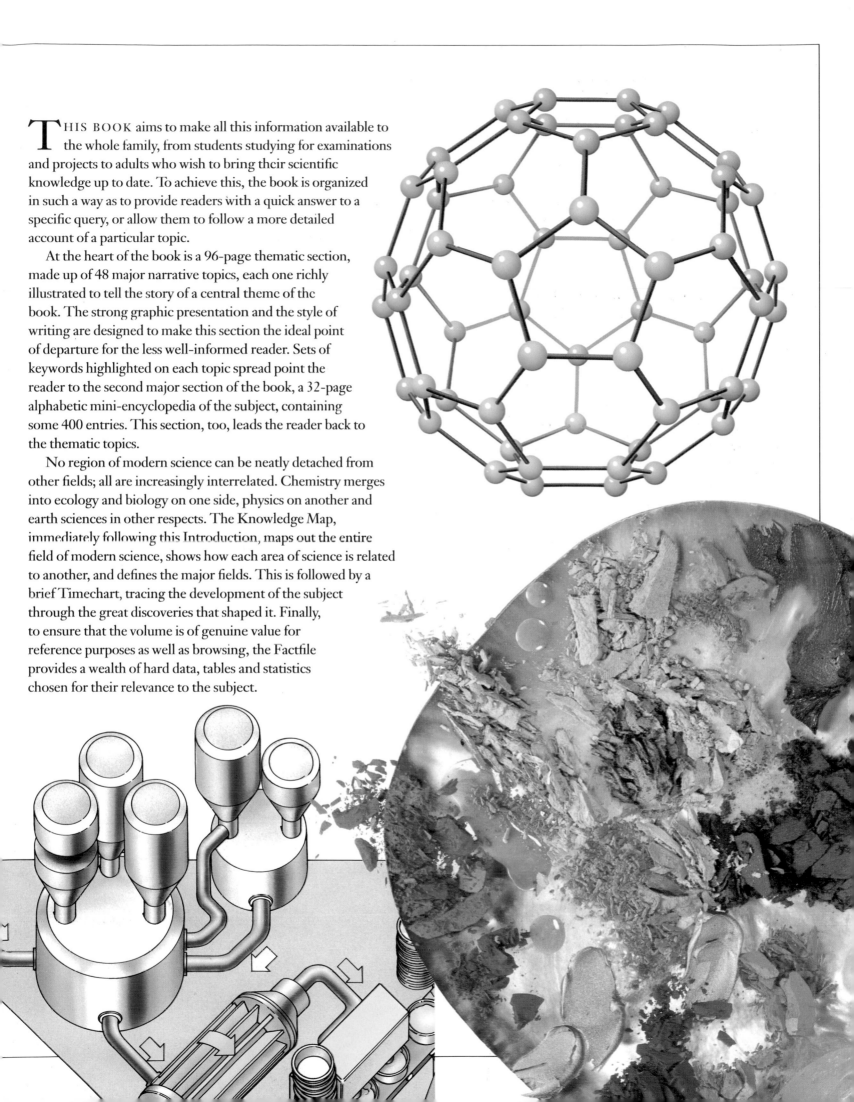

THIS BOOK aims to make all this information available to the whole family, from students studying for examinations and projects to adults who wish to bring their scientific knowledge up to date. To achieve this, the book is organized in such a way as to provide readers with a quick answer to a specific query, or allow them to follow a more detailed account of a particular topic.

At the heart of the book is a 96-page thematic section, made up of 48 major narrative topics, each one richly illustrated to tell the story of a central theme of the book. The strong graphic presentation and the style of writing are designed to make this section the ideal point of departure for the less well-informed reader. Sets of keywords highlighted on each topic spread point the reader to the second major section of the book, a 32-page alphabetic mini-encyclopedia of the subject, containing some 400 entries. This section, too, leads the reader back to the thematic topics.

No region of modern science can be neatly detached from other fields; all are increasingly interrelated. Chemistry merges into ecology and biology on one side, physics on another and earth sciences in other respects. The Knowledge Map, immediately following this Introduction, maps out the entire field of modern science, shows how each area of science is related to another, and defines the major fields. This is followed by a brief Timechart, tracing the development of the subject through the great discoveries that shaped it. Finally, to ensure that the volume is of genuine value for reference purposes as well as browsing, the Factfile provides a wealth of hard data, tables and statistics chosen for their relevance to the subject.

KNOWLEDGE MAP
Key Fields of Modern Science

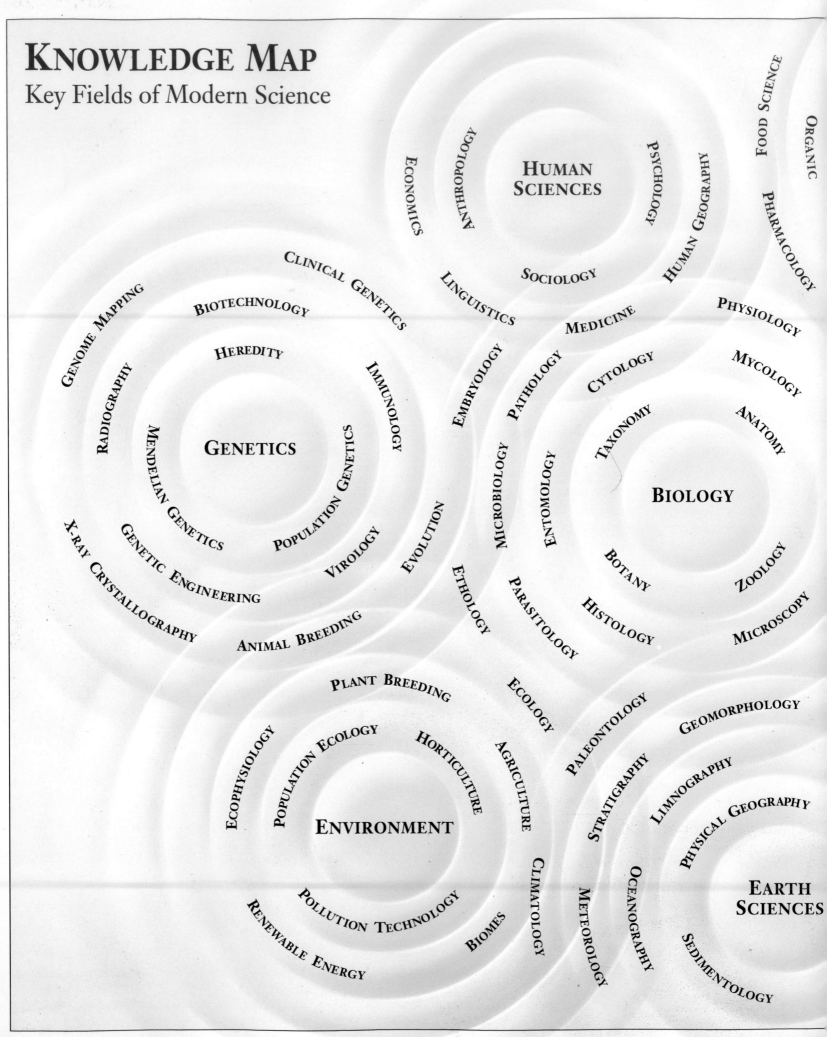

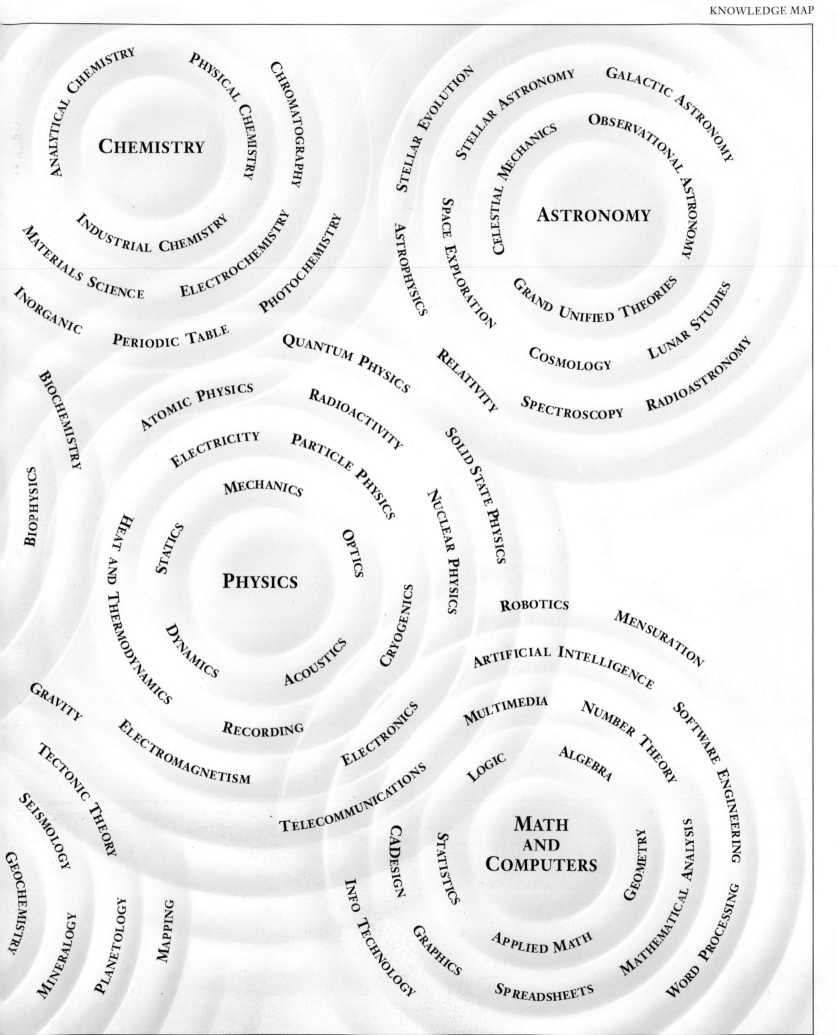

KNOWLEDGE MAP
Modern Chemistry

FOOD SCIENCE

The chemical study of food. Food scientists deal with many issues, from dietary requirements and the nutritional value of foods, to preparation and preservation. They are also concerned with the development of processed foods and chemicals to flavor, preserve and color food.

MATERIALS SCIENCE

Materials scientists are concerned with the physical and chemical properties of known materials and with developing new ones. The properties of these new materials are often engineered to make them suitable for specialized tasks.

ORGANIC CHEMISTRY

The study of the very large class of naturally occurring or synthesized compounds that contain carbon, often in association with hydrogen. Nitrogen, oxygen, sulfur, phosphorus and the halogens are other common elements. In organic compounds, most elements are bound to carbon by covalent bonds.

CHEMICAL ENGINEERING

The design, construction and operation of the plant and associated machines used in industrial chemistry. Containers and pipework, with pumps or hoppers, must tolerate a wide range of chemicals, from corrosive acids and alkalis, to poisonous gases and volatile solvents. Often a chemical plant has to operate at high temperatures and pressures.

INORGANIC CHEMISTRY

The study of naturally occurring or synthesized compounds that generally do not contain carbon, although a few simple carbon compounds such as the metallic carbonates are considered to be inorganic. In general, inorganic compounds are held together by ionic bonds. Inorganic chemistry is descriptive, synthetic and physical.

PHARMACOLOGY

The study of drugs and their effects. Pharmacologists study the sources, composition and preparation of medicinal substances. The field also includes therapeutics, the study of the use of drugs to treat disease. The development of new drugs may be based on a biochemical study of body reactions and computer design of molecules for specific reactions. The related subject, pharmacy, deals with drug preparation and storage.

INDUSTRIAL CHEMISTRY

The application of other branches of chemistry for use in industrial processes. This often involves the extraction of usable chemicals from raw materials, and the scaling up of laboratory reactions for use in large-scale production, using either continuous or batch processes, as well as devising practical methods of combining a series of chemical reactions that use simple raw materials to produce a desired product with a minimum energy input and waste.

ELECTRO-CHEMISTRY

The study of the chemical properties and reactions involving ions, usually in solution. Included in electrochemistry is the study of electrolysis, which occurs when an electric current is passed through an electrolyte containing a substance that conducts electricity when melted or dissolved.

COLLOID CHEMISTRY

The study of the preparation and properties of colloids, in which a dispersed phase of particles is distributed in a continuous phase – such as a sol. The particles are much larger than molecules but too small to be seen with an ordinary microscope.

ALCHEMY

An ancient pseudoscience in which chemical experiments were used to search for such mystical substances as the elixir of life. Alchemists did make some useful discoveries, giving rise to the science of chemistry.

PHYSICAL CHEMISTRY

The study of the effect of chemical structure on the physical properties of substances. Physical chemists apply the principles of physics to the study of chemical behavior. Included in their studies are topics such as chemical thermodynamics – the effect of changes in temperature and pressure on reactions – and electrochemistry.

NUCLEAR CHEMISTRY

The study of the chemistry of uranium and the transuranic elements, all of which have been synthesized in particle accelerators or recovered from the products of nuclear explosions. Often such elements are radioactive.

ANALYTICAL CHEMISTRY

The branch of chemistry concerned with the identification of substances qualitatively and quantitatively. Analytical chemists use a range of chemical and physical methods to help identify the components in a compound or mixture.

SPECTROSCOPY

The production and analysis of spectra of electromagnetic radiation absorbed or emitted by a sample. Spectroscopy involves techniques which monitor the behavior of electrons under different conditions. All provide information about chemical composition and physical properties of the elements or molecules in the sample.

CHROMATOGRAPHY

A technique for analyzing or separating mixtures of gases, liquids or dissolved substances by their molecular weight. The many techniques include paper chromatography, thin-layer and gas chromatography. All involve injecting the sample into a moving phase, passing it over a stationary phase and detecting the differential rates of adsorption of the various components of the mixture.

THEORETICAL CHEMISTRY

The study of elements at the atomic level to account for their properties and behavior. Of particular interest is the way in which electrons interact to form chemical bonds.

PERIODIC TABLE

A table of elements arranged in order of increasing atomic numbers to show the similarities of elements with related electron configurations. The Table also shows the chemical relationships between elements with different configurations. The vertical columns of the Table are known as groups, the horizontal rows are periods.

BIOCHEMISTRY

The study of the chemistry of living organisms, especially the structure of their chemical components and the reactions in which they are involved. Biochemistry forms an important part of physiology, nutrition and genetics, and the work of biochemists has an important impact in medicine, agriculture and industry.

PHOTOCHEMISTRY

The branch of chemistry concerned with photochemical reactions, which are those reactions brought about by the action of light or ultraviolet radiation. The primary step is the absorption of light energy by a particular atom, molecule or ion, which excites the particle and brings about a chemical reaction. Photosynthesis and the processes used in photography are photochemical reactions.

TIMECHART

ALCHEMY IS OFTEN REGARDED as the basis for modern chemistry. Alchemists tended not to theorize but to hone practical skills in the search for the transmutation of matter (the production of gold from base metals) and the elixir of life. The Chinese discovered gunpowder in this quest.

The medieval Arabs blended their pharmacological tradition with Greek, Indian and Chinese work to produce systematic surveys of alchemy. They also improved techniques such as distillation, crystallization and sublimation. In medieval Europe

alchemy was heretical and advances came from lone practitioners such as Paracelsus (1493–1541). By using mineral medicines he began to create chemistry as a reputable discipline.

Robert Boyle (1627–91) united early chemistry and physics with his study of gases. However, Newtonian advances in physics were not yet matched in chemistry. For example, the notion that "phlogiston" was ejected when substances burned endured until accurate measurement of the combustion products of metals showed that they gained, rather than lost, weight during the

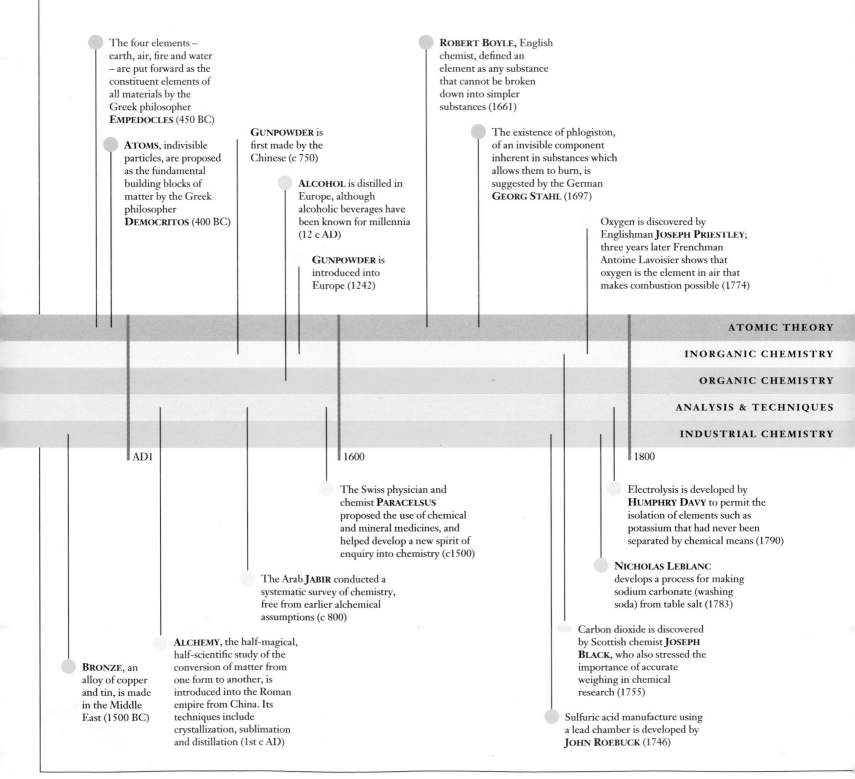

The four elements – earth, air, fire and water – are put forward as the constituent elements of all materials by the Greek philosopher **EMPEDOCLES** (450 BC)

ATOMS, indivisible particles, are proposed as the fundamental building blocks of matter by the Greek philosopher **DEMOCRITOS** (400 BC)

GUNPOWDER is first made by the Chinese (c 750)

ALCOHOL is distilled in Europe, although alcoholic beverages have been known for millennia (12 c AD)

GUNPOWDER is introduced into Europe (1242)

ROBERT BOYLE, English chemist, defined an element as any substance that cannot be broken down into simpler substances (1661)

The existence of phlogiston, of an invisible component inherent in substances which allows them to burn, is suggested by the German **GEORG STAHL** (1697)

Oxygen is discovered by Englishman **JOSEPH PRIESTLEY**; three years later Frenchman Antoine Lavoisier shows that oxygen is the element in air that makes combustion possible (1774)

ATOMIC THEORY

INORGANIC CHEMISTRY

ORGANIC CHEMISTRY

ANALYSIS & TECHNIQUES

INDUSTRIAL CHEMISTRY

AD1

1600

1800

The Swiss physician and chemist **PARACELSUS** proposed the use of chemical and mineral medicines, and helped develop a new spirit of enquiry into chemistry (c1500)

The Arab **JABIR** conducted a systematic survey of chemistry, free from earlier alchemical assumptions (c 800)

BRONZE, an alloy of copper and tin, is made in the Middle East (1500 BC)

ALCHEMY, the half-magical, half-scientific study of the conversion of matter from one form to another, is introduced into the Roman empire from China. Its techniques include crystallization, sublimation and distillation (1st c AD)

Electrolysis is developed by **HUMPHRY DAVY** to permit the isolation of elements such as potassium that had never been separated by chemical means (1790)

NICHOLAS LEBLANC develops a process for making sodium carbonate (washing soda) from table salt (1783)

Carbon dioxide is discovered by Scottish chemist **JOSEPH BLACK**, who also stressed the importance of accurate weighing in chemical research (1755)

Sulfuric acid manufacture using a lead chamber is developed by **JOHN ROEBUCK** (1746)

process. Antoine Lavoisier (1743–94), thought of as the father of chemistry, explained this as the reaction of oxygen (isolated by Joseph Priestley) in the air with the metal. Lavoisier also studied air as a chemically interactive substance and laid down the modern definition of an element.

The next step came with John Dalton (1766–1844). He clarified the notion of the atom and quantified it by experimental measurement, introducing the concept of relative atomic mass. He also gave each element a different symbol, an idea used by Jons

Jacob Berzelius (1779–1848) to produce the notation used today.

Electrochemistry gave impetus to analytical chemistry in the early 19th century. With it Humphry Davy (1778–1829) was able to break down substances then thought to be elements, and isolated sodium, potassium, calcium, barium and iodine. Another key factor was the adoption of methodical and reproducible techniques by scientists such as Michael Faraday (1791–1867). The rise of analytical chemistry had two distinct results: the growth of the chemical industry, and the idea of periodicity.

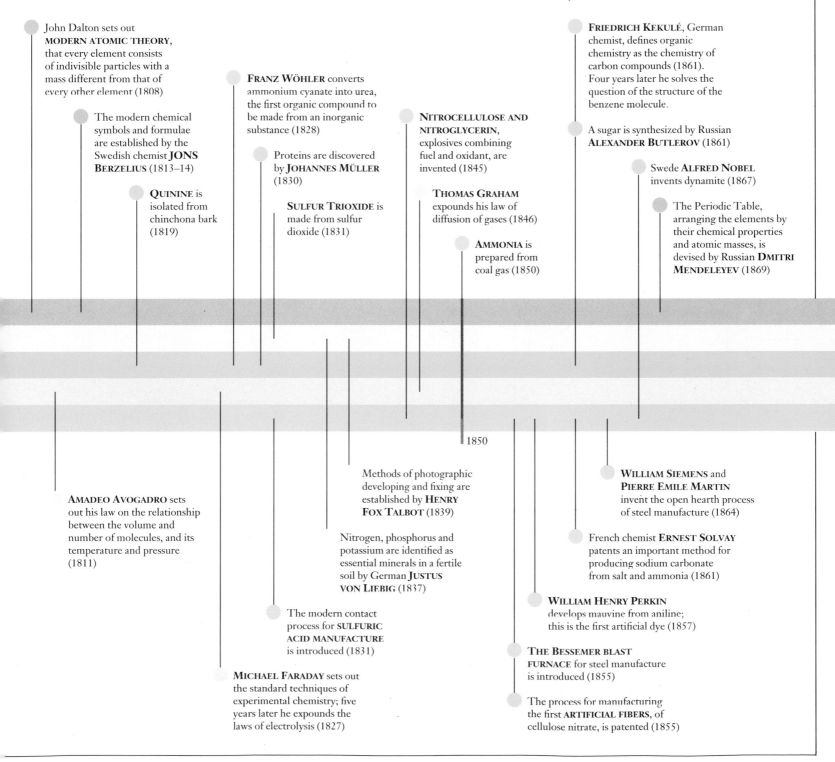

John Dalton sets out **MODERN ATOMIC THEORY,** that every element consists of indivisible particles with a mass different from that of every other element (1808)

The modern chemical symbols and formulae are established by the Swedish chemist **JONS BERZELIUS** (1813–14)

QUININE is isolated from chinchona bark (1819)

FRANZ WÖHLER converts ammonium cyanate into urea, the first organic compound to be made from an inorganic substance (1828)

Proteins are discovered by **JOHANNES MÜLLER** (1830)

SULFUR TRIOXIDE is made from sulfur dioxide (1831)

NITROCELLULOSE AND NITROGLYCERIN, explosives combining fuel and oxidant, are invented (1845)

THOMAS GRAHAM expounds his law of diffusion of gases (1846)

AMMONIA is prepared from coal gas (1850)

FRIEDRICH KEKULÉ, German chemist, defines organic chemistry as the chemistry of carbon compounds (1861). Four years later he solves the question of the structure of the benzene molecule.

A sugar is synthesized by Russian **ALEXANDER BUTLEROV** (1861)

Swede **ALFRED NOBEL** invents dynamite (1867)

The Periodic Table, arranging the elements by their chemical properties and atomic masses, is devised by Russian **DMITRI MENDELEYEV** (1869)

1850

AMADEO AVOGADRO sets out his law on the relationship between the volume and number of molecules, and its temperature and pressure (1811)

Methods of photographic developing and fixing are established by **HENRY FOX TALBOT** (1839)

Nitrogen, phosphorus and potassium are identified as essential minerals in a fertile soil by German **JUSTUS VON LIEBIG** (1837)

The modern contact process for **SULFURIC ACID MANUFACTURE** is introduced (1831)

MICHAEL FARADAY sets out the standard techniques of experimental chemistry; five years later he expounds the laws of electrolysis (1827)

WILLIAM SIEMENS and **PIERRE EMILE MARTIN** invent the open hearth process of steel manufacture (1864)

French chemist **ERNEST SOLVAY** patents an important method for producing sodium carbonate from salt and ammonia (1861)

WILLIAM HENRY PERKIN develops mauvine from aniline; this is the first artificial dye (1857)

THE BESSEMER BLAST FURNACE for steel manufacture is introduced (1855)

The process for manufacturing the first **ARTIFICIAL FIBERS,** of cellulose nitrate, is patented (1855)

In 1790 Nicholas Leblanc (1742–1806) had developed a process for preparing sodium carbonate, an important raw material in glass, paper, soap and textile manufacture. This led to the development of processes for other chemicals, such as sulfuric acid. William Henry Perkin's (1838–1907) discovery of the first artificial dye also encouraged the growth of a science-based chemical industry. One factor in this growth was the experimental technique of Justus von Leibig (1803–73), who searched for the perfect chemical method; the notion was passed on to industry.

The rapidly growing chemical industry required not only the scaling up of laboratory reactions but also a knowledge of the effects of temperature and pressure. The late 19th century saw the emergence of the new field of physical chemistry, when Jacobus Henricus van't Hoff (1852–1911) studied thermodynamics and chemical kinetics. From this work the ideas of mathematics were introduced into chemistry. More recently, physical chemistry has been reabsorbed into the broad modern fields of organic and inorganic chemistry.

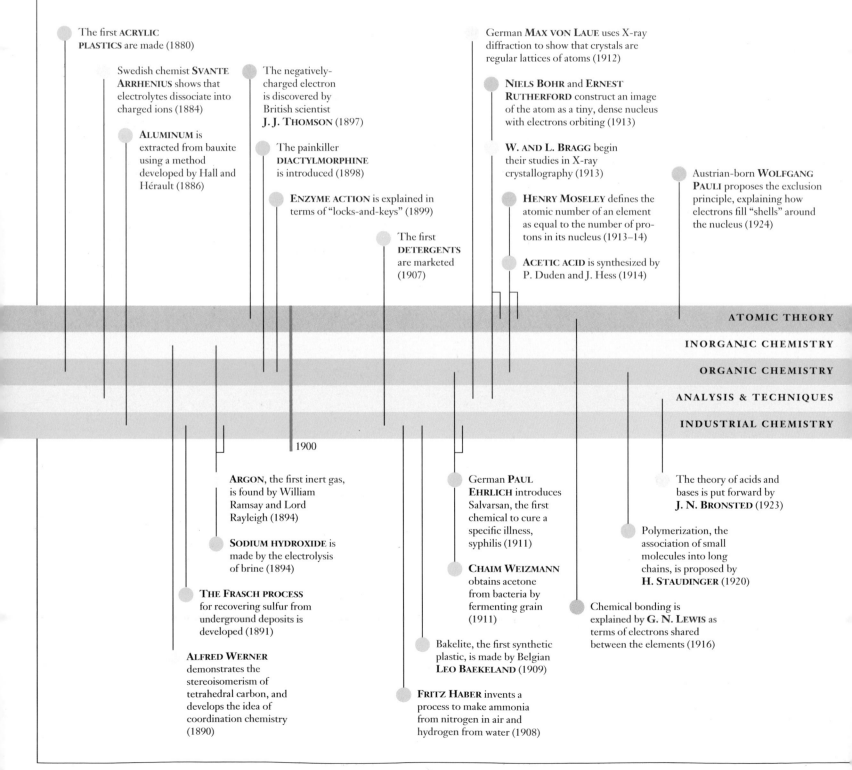

The first **ACRYLIC PLASTICS** are made (1880)

Swedish chemist **SVANTE ARRHENIUS** shows that electrolytes dissociate into charged ions (1884)

ALUMINUM is extracted from bauxite using a method developed by Hall and Hérault (1886)

The negatively-charged electron is discovered by British scientist **J. J. THOMSON** (1897)

The painkiller **DIACTYLMORPHINE** is introduced (1898)

ENZYME ACTION is explained in terms of "locks-and-keys" (1899)

The first **DETERGENTS** are marketed (1907)

German **MAX VON LAUE** uses X-ray diffraction to show that crystals are regular lattices of atoms (1912)

NIELS BOHR and **ERNEST RUTHERFORD** construct an image of the atom as a tiny, dense nucleus with electrons orbiting (1913)

W. AND L. BRAGG begin their studies in X-ray crystallography (1913)

HENRY MOSELEY defines the atomic number of an element as equal to the number of protons in its nucleus (1913–14)

ACETIC ACID is synthesized by P. Duden and J. Hess (1914)

Austrian-born **WOLFGANG PAULI** proposes the exclusion principle, explaining how electrons fill "shells" around the nucleus (1924)

ATOMIC THEORY

INORGANIC CHEMISTRY

ORGANIC CHEMISTRY

ANALYSIS & TECHNIQUES

INDUSTRIAL CHEMISTRY

1900

ARGON, the first inert gas, is found by William Ramsay and Lord Rayleigh (1894)

SODIUM HYDROXIDE is made by the electrolysis of brine (1894)

THE FRASCH PROCESS for recovering sulfur from underground deposits is developed (1891)

ALFRED WERNER demonstrates the stereoisomerism of tetrahedral carbon, and develops the idea of coordination chemistry (1890)

German **PAUL EHRLICH** introduces Salvarsan, the first chemical to cure a specific illness, syphilis (1911)

CHAIM WEIZMANN obtains acetone from bacteria by fermenting grain (1911)

Bakelite, the first synthetic plastic, is made by Belgian **LEO BAEKELAND** (1909)

FRITZ HABER invents a process to make ammonia from nitrogen in air and hydrogen from water (1908)

The theory of acids and bases is put forward by **J. N. BRONSTED** (1923)

Polymerization, the association of small molecules into long chains, is proposed by **H. STAUDINGER** (1920)

Chemical bonding is explained by **G. N. LEWIS** as terms of electrons shared between the elements (1916)

The second advance from the use of analytical techniques was made by Dmitri Mendeleyev (1834–1907) with the concept of periodicity. He arranged the elements according to their chemical properties and atomic weights. He was even able to predict the existence of undiscovered elements. His Periodic Table also helped physicists to show the atom as made up of electrons surrounding a small, heavy nucleus. This led Gilbert N. Lewis (1875–1946) to explain chemical bonding in terms of valence and the sharing of electrons. Linus Pauling (1901-94) introduced quantum mechanics into the understanding of the chemical bond; Pauling's valence bond was later superseded by the molecular orbital approach. Knowledge of chemical bonds and reaction mechanisms led to great progress in products such as drugs and plastics, while X-ray techniques unraveled the structure of complex organic molecules and allowed them to be synthesized.

By the 1990s few fundamental challenges remained, though the chemist's skills were required in increasingly diverse areas, including superconductivity and conducting polymers.

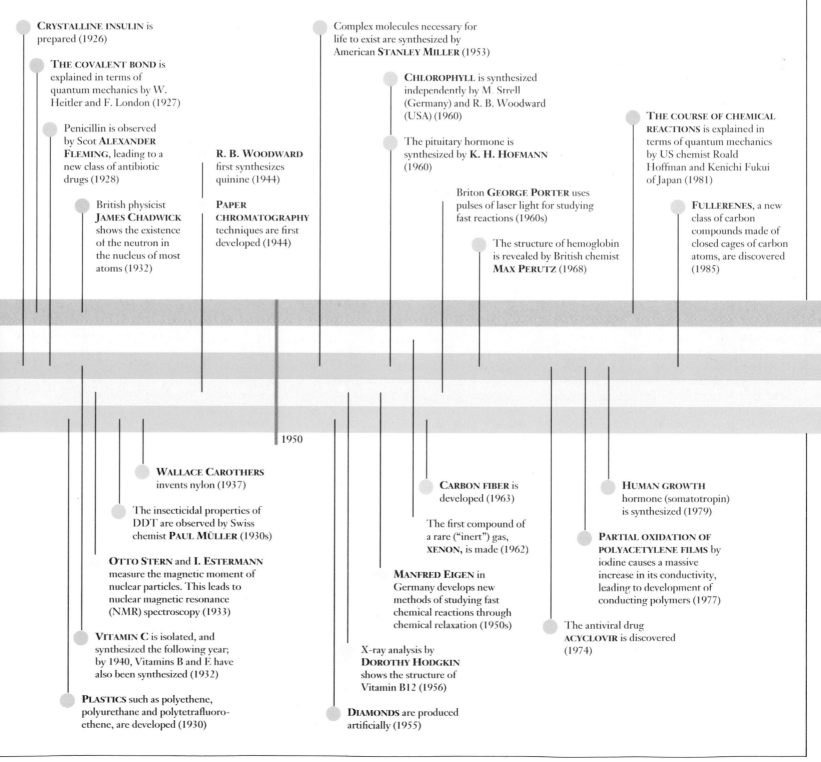

CRYSTALLINE INSULIN is prepared (1926)

THE COVALENT BOND is explained in terms of quantum mechanics by W. Heitler and F. London (1927)

Penicillin is observed by Scot **ALEXANDER FLEMING**, leading to a new class of antibiotic drugs (1928)

British physicist **JAMES CHADWICK** shows the existence of the neutron in the nucleus of most atoms (1932)

R. B. WOODWARD first synthesizes quinine (1944)

PAPER CHROMATOGRAPHY techniques are first developed (1944)

Complex molecules necessary for life to exist are synthesized by American **STANLEY MILLER** (1953)

CHLOROPHYLL is synthesized independently by M. Strell (Germany) and R. B. Woodward (USA) (1960)

The pituitary hormone is synthesized by **K. H. HOFMANN** (1960)

Briton **GEORGE PORTER** uses pulses of laser light for studying fast reactions (1960s)

The structure of hemoglobin is revealed by British chemist **MAX PERUTZ** (1968)

THE COURSE OF CHEMICAL REACTIONS is explained in terms of quantum mechanics by US chemist Roald Hoffman and Kenichi Fukui of Japan (1981)

FULLERENES, a new class of carbon compounds made of closed cages of carbon atoms, are discovered (1985)

1950

WALLACE CAROTHERS invents nylon (1937)

The insecticidal properties of DDT are observed by Swiss chemist **PAUL MÜLLER** (1930s)

OTTO STERN and **I. ESTERMANN** measure the magnetic moment of nuclear particles. This leads to nuclear magnetic resonance (NMR) spectroscopy (1933)

VITAMIN C is isolated, and synthesized the following year; by 1940, Vitamins B and E have also been synthesized (1932)

PLASTICS such as polyethene, polyurethane and polytetrafluoro-ethene, are developed (1930)

CARBON FIBER is developed (1963)

The first compound of a rare ("inert") gas, **XENON,** is made (1962)

MANFRED EIGEN in Germany develops new methods of studying fast chemical reactions through chemical relaxation (1950s)

X-ray analysis by **DOROTHY HODGKIN** shows the structure of Vitamin B12 (1956)

DIAMONDS are produced artificially (1955)

HUMAN GROWTH hormone (somatotropin) is synthesized (1979)

PARTIAL OXIDATION OF POLYACETYLENE FILMS by iodine causes a massive increase in its conductivity, leading to development of conducting polymers (1977)

The antiviral drug **ACYCLOVIR** is discovered (1974)

Chemistry
KEYWORDS

acid

A compound that gives rise to hydrogen ions (H^+ or protons) when in solution in an ionizing solvent (usually water). Acids are proton donors and accept electrons to form ionic bonds. They react with bases to form salts and also act as solvents. Strong acids are corrosive; dilute acids have a sour or sharp taste. Acids can be detected by using colored indicators and have a pH value of less than 7. They can be classified according to their basicity (the number of hydrogen atoms available to react with a base) and degree of ionization (how many of the available hydrogen atoms dissociate in water). Dilute sulfuric acid is a strong (highly ionized) dibasic acid. Most naturally occurring acids are found as organic compounds, such as the fatty acids (R.COOH) and sulfonic acids (R.SO$_3$H).

> ### CONNECTIONS
>
> ACIDS, BASES AND SALTS 62
> A CLEANER ENVIRONMENT 70
> MAKING ACIDS AND BASES 76
> LIVING CHEMISTRY 112
> PHOTOGRAPHY 128

acid rain

Unnaturally acidic rainfall, caused principally by the release into the atmosphere of sulfur dioxide (SO_2) and oxides of nitrogen. Sulfur dioxide is formed from the burning of fossil fuels that contain high quantities of sulfur; nitrogen oxides are produced by industrial activities and from automobile exhaust fumes (*see* **exhaust emission** and **catalytic converter**).

> ### CONNECTIONS
>
> A CLEANER ENVIRONMENT 70
> ATMOSPHERIC CHEMISTRY 132

acrylic

A synthetic resin produced by the polymerization of esters or other derivatives of acrylic acid (propenoic acid). Acrylics are usually transparent and thermoplastic, and resistant to light, aging and a number of chemical agents. Acrylic resins are used widely for lenses and instrument covers. Examples include polypropanonitrile and polymethyl methacrylate.

actinide

Any of a series of 15 radioactive metallic elements in the **Periodic Table** of elements ranging from atomic number 89 (actinium) to 103 (lawrencium). Elements 89 to 95 occur naturally; the rest are synthesized (*see* **transuranium element**). Actinides are grouped together because of their chemical similarities; their properties differ only slightly with increasing atomic number. *See also* **lanthanide**.

activation energy

The minimum energy required in order to start a chemical reaction. Some elements and compounds react merely by being brought into contact (*see* **spontaneous combustion**). For others it is necessary to supply energy to start the reaction. The activation energy is expressed as joules per mole of reactants.

active site

1 The site on the surface of a catalyst where the activity occurs. **2** The site on the surface of an enzyme molecule that binds the substrate molecule. The interactions that occur between the enzyme and substrate molecule depend on the three-dimensional arrangement of the polypeptide chains of the enzyme molecule. This arrangement is responsible for the specificity of the enzyme and for any susceptibility to inhibition.

addition

A reaction process in which the atoms of an element or compound react with a double bond or triple bond in an organic compound by opening up one of the bonds and becoming attached to it. One compound is thus added to another compound; for example $CH_2{=}CH_2 + HCl \rightarrow CH_3CH_2Cl$.

An example is the addition of hydrogen atoms to unsaturated compounds in vegetable oils to produce margarine (**hydrogenation**). Addition-elimination reactions are addition reactions formed by the elimination of another molecule. *See also* **substitution**.

adhesive

Any substance used for joining two surfaces together. Natural adhesives include gelatin and vegetable gums. Synthetic adhesives include **thermoplastic** and **thermosetting resins** (which are often stronger than the substances they join), mixtures of epoxy resin and hardener that set by chemical reaction, and elastomeric adhesives for flexible joints.

alcohol

A member of a group of organic chemical compounds that contain one or more hydroxyl (–OH) groups in the molecule and that form esters with acids. It may be liquid or solid, according to the size and complexity of the molecule. The five simplest alcohols form a **homologous series** in which the number of carbon and hydrogen atoms increases progressively. Each one has an additional CH_2 (methylene) group in the molecule: methanol (CH_3OH), ethanol (C_2H_5OH), propanol (C_3H_7OH), butanol (C_4H_9OH) and pentanol ($C_5H_{11}OH$). Alcohols containing the –CH_2OH group are primary; those containing –CHOH are secondary; while those containing –COH are tertiary. Alcohols are used as solvents, in making dyes, in perfumery, and in pharmacology. Ethanol is produced naturally by fermentation and has an intoxicating effect.

> ### CONNECTIONS
>
> MAKING HYDROCARBONS 92
>
> TESTING FOR DRUGS 118
>
> DYES AND DYEING 124

aldehyde

A member of a group of organic compounds that may be prepared by oxidation of primary alcohols, so that the hydroxyl (–OH) group loses its hydrogen to give an oxygen joined by a double bond to a carbon atom (the aldehyde group, –CHO). The name is derived from **alcohol dehyd**rogenation. Aldehydes are usually liquids and include methanal (formaldehyde), ethanal (acetaldehyde) and benzaldehyde. *See also* **ketone**.

alicyclic compound

See **cyclic compound**.

aliphatic compound

An organic compound in which the carbon atoms are joined in straight chains (hexane, C_6H_{14}), or in branched chains (2-methylpentane, $CH_3CH(CH_3)CH_2CH_2CH_3$). Aliphatic compounds have bonding electrons localized within the vicinity of the bonded atoms. Cyclic compounds that do not have delocalized electrons are also aliphatic, as in the alicyclic compound cyclohexane (C_6H_{12}) or the heterocyclic piperidine ($C_5H_{11}N$). *See also* **aromatic compound**.

alkali

A compound that is classed as a base and dissolves in water to give hydroxide ions (OH^-). Alkalis have a pH of more than 7. They react with acids to form a salt and water (neutralization). The four main alkalis are sodium hydroxide (NaOH), potassium hydroxide (KOH), calcium hydroxide ($Ca(OH)_2$) and aqueous ammonia (NH_4OH).

alkali metal

Any of lithium, sodium, potassium, rubidium, cesium and francium, which form Group I in the **Periodic Table** of the elements. Alkali metals have a valence of one and have a very low density; lithium, sodium and potassium float on water. In general they are soft metals with high reactivity and a low melting point. Because of their reactivity, alkali metals are only found as compounds in nature (usually as salts) and are usually used as chemical reactants rather than as structural metals.

alkaline earth metal

Any of the elements beryllium, magnesium, calcium, strontium, barium and radium, which form Group II in the **Periodic Table** of the elements. Alkaline earth metals are strong bases, have a valence of two and occur in nature only in compounds. They and their compounds are used to make alloys, oxidizers and drying agents.

alkaloid

Any of a group of physiologically active nitrogenous, organic compounds derived from plants. They form salts with acids and, when soluble, give alkaline solutions. Substances in this group are included by custom rather than by scientific rules. Alkaloids have diverse pharmacological properties; examples include morphine, cocaine, quinine, caffeine, strychnine, nicotine and atropine.

ALLOTROPE

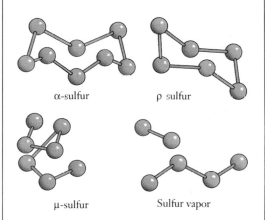

α-sulfur

ρ-sulfur

μ-sulfur

Sulfur vapor

alkane

Any hydrocarbon with the general formula C_nH_{2n+2}, traditionally known as a **paraffin**. Lighter alkanes, such as methane, ethane, propane and butane, are colorless gases; heavier ones are liquids or solids. They are found in natural gas and petroleum. All alkanes are saturated compounds: that is, they contain only single covalent bonds.

alkene

Any hydrocarbon that has the general formula C_nH_{2n}. Alkenes were formerly known as **olefins**. Lighter alkenes, such as ethene and propene, are gases, obtained from the cracking of oil fractions. Alkenes are unsaturated compounds with one or more double bonds between adjacent carbon atoms.

> ### CONNECTIONS
>
> COMBUSTION AND FUEL 66
>
> HYDROCARBON CHAINS 84
>
> CARBON-HYDROGEN COMPOUNDS 86

alkyl group

Any chemical group derived by removing a hydrogen atom from an alkane, usually denoted by the symbol R; for example, the methyl group, –CH_3.

alkyne

Any hydrocarbon that has one or more **triple bonds** between adjacent carbon atoms. A simple example is ethyne (acetylene, CH≡CH). Alkynes polymerize to form **aromatic compounds**.

allotrope

One of two or more forms of an element in which each possesses different physical properties but is the same state of matter (gas, liquid or solid). Oxygen exists as two gaseous allotropes: "normal" oxygen (O_2) and ozone (O_3), which differ in their molecular configurations. The allotropes of carbon are diamond, graphite and the molecule C_{60} (buckminsterfullerene). Sulfur, tin and phosphorus also show allotropy.

alloy

Any material made of a blend of a metal with another substance to give it special qualities, such as resistance to corrosion or greater hardness. Common alloys include bronze, brass, cupronickel, German silver, gunmetal, pewter, solder, steel and stainless steel.

aluminum

A lightweight, ductile and malleable metallic element. Its symbol is Al, its atomic number 13, and its relative atomic mass 26.9815.

Aluminum is the third most abundant element (and the most abundant metal) in the Earth's crust. It is an excellent conductor of electricity and oxidizes easily; it is also highly resistant to tarnish. It is prepared commercially by the electrolysis of bauxite (*see* **Hall–Héroult process**). When combined with copper, silicon or magnesium it forms **alloys** of great strength. It is widely used in the construction of ships, aircraft and space vehicles.

amalgam

An **alloy** of mercury with one or more metals. Most metals form amalgams, except iron and platinum. Amalgam is used in dentistry for filling teeth, and usually contains copper, silver and zinc as the main alloying ingredients. **Amalgamation**, the process of forming an amalgam, is a technique sometimes used to extract gold and silver from their ores. The ores are treated with mercury, which combines with the precious metals.

amide

Any of a class of organic compounds derived from a fatty acid by replacing the hydroxyl group (–OH) with an amino group (–NH$_2$) to produce the amide group (–CONH$_2$). One of the simplest amides is ethanamide (acetamide, CH$_3$CONH$_2$). Amides can be produced by heating the ammonium salt of the corresponding carboxylic acid.

amine

Any of a class of organic compounds in which one or more of the hydrogen atoms of ammonia (NH$_3$) have been replaced by alkyl groups. They can be classified according to the number of hydrogen atoms replaced in ammonia (that is, primary, secondary and tertiary amines). Amines are colorless gases or liquids. **Aromatic** amine compounds include aniline, which is used in dyeing.

amino acid

Any of a group of water-soluble organic compounds, mainly composed of carbon, oxygen, hydrogen and nitrogen, that contain both a basic amino group (–NH$_2$) and an acidic carboxyl (–COOH) group. Two or more amino acids are joined together are known as peptides. Proteins are made up of interacting polypeptides (peptide chains consisting of more than three amino acids) and are folded or twisted.

ammonia

A colorless pungent-smelling gas (NH$_3$), which is lighter than air and dissolves in water to give a basic solution. It is made on an industrial scale by the **Haber process**, and used to produce nitric acid, nitrogenous fertilizers and explosives. Ammonia plays an important role in the nitrogen cycle; a number of aquatic organisms and insects excrete nitrogenous waste in the form of ammonia.

ammonia–soda process

A multi-stage industrial process for the manufacture of sodium carbonate. Carbon dioxide is first generated by heating limestone and passed through a solution of sodium chloride saturated with ammonia. Sodium hydrogen carbonate is then isolated and heated to yield sodium carbonate. All intermediate byproducts are recycled so that the only ultimate byproduct is calcium chloride. The process is also known as the **Solvay process** after its inventor.

anhydride

Any compound that is obtained by the removal of water from another compound, usually a dehydrated acid. For example, sulfur trioxide is the anhydride of sulfuric acid: H$_2$SO$_4$ → H$_2$O + SO$_3$.

anion

A negatively charged ion. In an electrolytic cell, a salt – for example, sodium chloride (NaCl) – is dissociated into positive Na$^+$ **cations** and negative charge Cl$^-$ anions, which are attracted to the positive electrode (*see* **anode**).

anode

A positive electrode. In an electrolytic cell it is the electrode toward which negative particles (anions) are attracted. The anode of an electrolytic cell obtains its positive charge by the application of an external electrical potential. *See also* **electrolysis**.

anodizing

A process that increases the corrosion resistance of a metal by building up a protective oxide layer on the surface of the metal. The metal is made the anode in an electrolytic cell containing an oxidizing electrolyte. The oxide layer produced can absorb dyes to produce decorative finishes.

aramid fibers

Synthetic fibers produced from linear copolymers containing repeating amide groups joined directly to two aromatic rings. The resulting fibers are very strong and are often used in composite materials. The term aramid is derived from **ar**omatic **amide**.

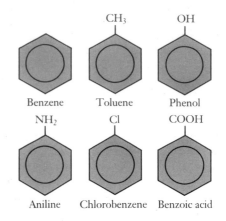

aromatic compound

Any organic compound, such as phenols, toluene, benzoic acid or salicylic acid, that incorporates a **benzene ring** in its structure (*see also* **cyclic compound**) or that has chemical properties similar to benzene. Although aromatic compounds are not saturated, they rarely undergo addition reactions; instead, they usually undergo electrophilic substitution reactions. The bonding electrons in aromatic compounds are delocalized (shared among several atoms within the molecule and not localized in the vicinity of the atoms involved in bonding).

Aromatic behavior is also found in heterocyclic compounds, such as pyridine, in which one of the carbon atoms of the benzene ring has been replaced by an atom of another element. Aromatic compounds react with concentrated nitric acid to form nitro derivatives and with concentrated sulfuric acid to form sulfonated derivatives.

aryl group

A group of chemical compounds derived from the removal of a hydrogen atom from an **aromatic compound**. For example, the phenyl group, C$_6$H$_5^-$, is obtained by the removal of hydrogen from benzene.

asbestos

Any of a number of related minerals with a fibrous structure that offer great heat resistance because of their non-flammability and poor conductivity. They are also chemically inert and have a high electrical resistance. Commercial asbestos is generally made from olivine, a serpentine mineral; tremolite (a white amphibole); and crocidolite (a blue amphibole). Although it is a valuable industrial mineral, the use of asbestos is now strictly controlled because exposure to its dust can cause the respiratory disease asbestosis. Most of the world's asbestos supply derives from Canada and Russia.

atom
The smallest part of an **element** that can take part in a chemical reaction and that cannot be broken down chemically into anything simpler. An atom is made up of **protons** and **neutrons** in a central nucleus surrounded by **electrons**. The electrons are arranged around the nucleus of an atom in distinct energy levels (*see* **orbital**). These shells can be regarded as a series of concentric spheres, each of which can contain a certain maximum number of electrons. The energy levels are usually numbered beginning with the shell nearest to the nucleus. The outermost shell is known as the valence shell because it contains the valence electrons. The lowest energy level, or innermost shell, can contain no more than two electrons. Outer shells are considered to be stable when they contain eight electrons but additional electrons can sometimes be accommodated provided that the outermost shell has a stable configuration. Electrons in unfilled shells are available to take part in chemical bonding, giving rise to the concept of valence. In ions, the electron shells contain more or fewer electrons than are required for a neutral atom, generating negative or positive charges. The atoms of the various elements differ in atomic number, relative atomic mass and chemical behavior. There are 106 different types of atom, corresponding with the 106 known elements as listed in the **Periodic Table** of the elements. *See also* **subatomic particle**.

CONNECTIONS

atomic mass
See **relative atomic mass (RAM)**.

atomic number
The number of protons in the nucleus of an atom, equal to the positive charge on the nucleus and equivalent to the number of electrons in a neutral atom. The elements of the Periodic Table are arranged according to their atomic number. From this it is

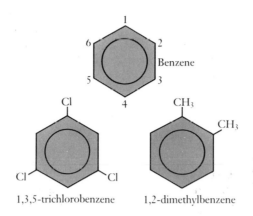

1,3,5-trichlorobenzene   1,2-dimethylbenzene

possible to deduce the electronic structure of an atom. For example, sodium has atomic number 11 and its electronic configuration is two electrons in the first energy level, eight electrons in the second energy level and one electron in the third energy level.

atomic weight
See **relative atomic mass (RAM)**.

Avogadro's constant
The number of atoms or molecules in one mole of a substance (6.02253×10^{23}); for example, the number of carbon atoms in 12 grams of the carbon 12 isotope. The **relative atomic mass** of any element, expressed in grams, contains this number of atoms. It is derived from a hypothesis first proposed by the Italian physicist Amedeo Avogadro.

azo dye
A group of synthetic dyes containing an azo group of two nitrogen atoms ($N=N$) connecting aromatic ring compounds, for example azobenzene. Azo dyes are usually red, brown or yellow. They are manufactured from aromatic amines.

balance
Any device for weighing small quantities of chemicals with a high degree of accuracy. The simple beam balance uses two pans suspended from a centrally pivoted beam. The material to be weighed is placed in one pan and known weights are placed in the other until the beam is horizontal. *See* **electronic balance**.

barbiturate
Any chemical derivative of barbituric acid ($CO(NHCO)_2CH_2$), an organic **heterocyclic compound**. Barbiturates are very important in pharmacology and are commonly used in medicine as sedatives or anesthetics.

base
A substance that is the complement of an acid; bases are defined as substances that accept hydrogen ions (H^+) or protons. Bases may contain negative ions such as the hydroxide ion (OH^-), which is the strongest base; or they may be molecules such as ammonia (NH_3). Ammonia is a weak base, as only some of its molecules accept protons. Many bases are insoluble, but bases that dissolve in water are called **alkalis**. Inorganic bases are usually oxides or hydroxides of metals, which react with dilute **acids** to form a salt and water. Many carbonates also react with dilute acids, giving off carbon dioxide.

benzene ring
The cyclic aromatic grouping (C_6H_6), consisting of a ring of six carbon atoms, all of which are in a single plane. The bonds between the carbon molecules are equivalent, and intermediate in length between single and double bonds. Benzene is the simplest **aromatic compound**.

bicarbonate
See **hydrogen carbonate**.

binder
Any substance added to another substance, particularly a paint or lacquer, to give cohesion. The binder may set at room temperature or may need to be heated before it has any permanent effect. Major groups of binders include phenolic resins, cellulose esters and ethers, and synthetic binders produced as spin-offs from the plastics industry.

biochemistry
The branch of chemistry concerned with living organisms, in particular the structure and reactions of proteins (such as enzymes), nucleic acids, carbohydrates and lipids. The study of biochemistry has increased knowledge of how animals and plants react with their environments; for example, in creating and storing energy by photosynthesis, taking in food and releasing waste products, and passing on their characteristics through their genes. Biochemistry plays an important part in many areas of current research, including medicine and agriculture.

CONNECTIONS

biodegradable plastic
Any plastic that can be broken down by bacterial enzymes. Most are based on material of biological origin such as starch and, more

KEYWORDS

recently, polymers of lactic acid. Most synthetic plastics cannot be broken down in this way. Because synthetic polymers, unlike metals, are expensive to recycle, the disposal of nonbiodegradable plastic packaging is a major environmental problem.

biotechnology

The application of living organisms to the industrial and commercial manufacture of food or other chemical products. The brewing and baking industries have long relied on yeast for fermentation purposes, while the dairy industries use a range of bacteria and fungi to convert milk into cheese and yogurts. Enzymes, whether extracted from cells or produced artificially, are central to the performance of chemical conversions in most biotechnological applications, being very cheap and efficient by comparison with the comparable chemical processes. The techniques of biotechnology are increasingly used in the fine chemical industry, in the production of flavors, perfumes and drugs.

blood test

A laboratory test to analyze blood composition. It may be used to detect drugs, alcohol or disease (*see* **diabetes**), and to determine the blood group of an individual.

blow molding

A process used to produce single-piece plastic items (such as bottles). It uses a combination of extrusion and blowing with a split mold that can be easily dismantled.

boiling point

The temperature at which the application of heat raises the temperature of a liquid no further, but converts it into vapor. The boiling point of water under normal pressure is 100°C. The lower the pressure, the lower the boiling point; the higher the pressure, the higher the boiling point.

bond

The result of the forces of attraction that hold together atoms to form molecules. The main types of bonding are ionic, covalent, metallic and intermolecular (such as hydrogen bonding). The type of chemical bond formed depends on the elements involved and their electronic structure.

In an **ionic** or **electrovalent bond**, common in inorganic compounds, the combining atoms gain or lose electrons to become ions; in sodium chloride (NaCl) sodium (Na) loses an electron to form a sodium ion (Na^+) while chlorine (Cl) gains an electron to form a chloride ion (Cl^-). In a **covalent bond**, the atomic orbitals of two atoms overlap to form a molecular orbital containing two electrons,

which are thus effectively shared between the two atoms. Covalent bonds are common in organic compounds, such as the four carbon–hydrogen bonds in methane (CH_4). In a **dative covalent** or **coordinate bond**, one of the combining atoms supplies both of the shared electrons in the bond.

Metallic bonds occur between metals in a crystal lattice; the atoms occupy lattice positions as positive ions and the valence electrons are shared between all the ions in an "electron gas". In a **hydrogen bond**, a hydrogen atom joined to an electronegative atom, such as nitrogen or oxygen, becomes partially positively charged and is weakly attracted to another electronegative atom on a neighboring molecule. *See also* **van der Waals' force**.

CONNECTIONS

TYPES OF BONDS 54

BONDS AND STRUCTURES 56

NAMES AND FORMULAS 60

breath test

An analytical test that determines the quantity of alcohol in the bloodstream. In one common breath test the individual breathes into a plastic bag connected to a tube containing an **indicator** (such as a diluted solution of potassium dichromate in 50 percent sulfuric acid) that changes color in the presence of alcohol. Another method is based on gas chromatography.

BOND

Water

Ethene

Hydroxide ion   OH⁻

Carbonate ion CO_3^{2-}

H_2O

C_2H_4

burning

See **combustion**.

calendering

The process used to impart a desired finish or to ensure uniform thickness of a textile. Calendering is usually achieved by passing the material through a series of special rollers under pressure.

carbide

A compound of carbon together with one other chemical element, usually a metal, silicon or boron. Calcium carbide (CaC_2) can be used as the starting material for the synthesis of many basic organic chemicals, by the addition of water and generation of ethyne. Some metallic carbides, such as tungsten carbide, are used in engineering due to their extreme hardness and strength.

carbohydrate

A compound composed of carbon, hydrogen and oxygen, with the basic formula $C_n(H_2O)_m$, and related compounds with the same basic structure but modified functional groups. As sugar and starch, carbohydrates form a major energy-providing part of the human diet. The simplest carbohydrates are sugars (**monosaccharides**, or single ring-shaped molecules, such as glucose and fructose, and **disaccharides**, double ring molecules such as sucrose), which are sweet-tasting soluble compounds. These basic sugar units may be joined together in long chains or branching structures to form **polysaccharides**, such as starch and glycogen. Even more complex carbohydrates include chitin, which is found in the cell walls of fungi and the hard outer skeletons of insects, and cellulose, which makes up the cell walls of plants.

CONNECTIONS

CHEMISTRY OF LIFE 108

VITAL RAW MATERIALS 110

LIVING CHEMISTRY 112

carbon

A non-metallic element. Its symbol is C, its atomic number 6, and its relative atomic mass 12.011. Carbon occurs on its own as the allotropes diamond and graphite; as fullerenes; as compounds in carbonaceous rocks such as chalk and limestone; as carbon dioxide in the atmosphere; as hydrocarbons in petroleum, coal and natural gas; and as a constituent of all organic substances. When added to iron, carbon forms a wide range of steel **alloys** with useful properties. *See also* **carbon–carbon bond**.

carbonate

A salt of carbonic acid (H_2CO_3), which forms the carbonate anion CO_3^{2-}. In nature, carbon dioxide (CO_2) dissolves sparingly in water (for example, when rain falls through the air) to form carbonic acid (H_2CO_3), which unites with various basic substances to form carbonates. Calcium carbonate ($CaCO_3$), found in chalk, limestone and marble, is one of the most abundant carbonates known. Seashells are also composed of calcium carbonate. Carbonates give off carbon dioxide when heated or treated with dilute acids. The acid reaction is used as the laboratory test for the presence of the ion, as it gives an immediate effervescence, with the gas turning a solution of calcium hydroxide (lime water) cloudy.

carbon–carbon bonds

A bond formed by carbon atoms joining with other carbon atoms. Carbon is unique among elements in the number of compounds it can form with other elements. It exhibits its valence in many ways, forming single, double and triple bonds with itself and with many other atoms. This is related to the great strength of the carbon–carbon bond, which means that carbon atoms can link indefinitely with each other to form molecules consisting of chains of straight and branched structures, or closed rings.

carbon cycle

The sequence of chemical reactions by which carbon circulates and is recycled through the ecosystem. Carbon from carbon dioxide is taken in by plants during photosynthesis and converted into carbohydrates, releasing the oxygen back into the atmosphere. The carbohydrates are then used in respiration by the plant, or by the animals that eat the plants, and are decomposed to release carbon dioxide back into the atmosphere. Carbon dioxide is also released into the atmosphere by the burning of fossil fuels. Today, the carbon cycle is in danger of being disrupted by the increased consumption and burning of fossil fuels and the burning of large tracts of tropical forests. As a result, levels of carbon dioxide are building up in the atmosphere and probably contributing to the **greenhouse effect**.

carbon dioxide

A colorless gas (CO_2), slightly soluble in water and denser than air, which is formed when carbon and carbon-containing compounds are fully oxidized. It is also produced when acids are added to carbonates or hydrogen carbonates and when these are heated. Like other acidic oxides, it dissolves in water to give a weak dibasic acid and forms salts with alkalis. With a solution of calcium hydroxide, CO_2 forms a white precipitate of calcium carbonate, which is used to test for its presence. The gas extinguishes most lower-temperature flames and is used in fire extinguishers. Carbon dioxide plays a vital role in the **carbon cycle**, in the formation of hard water and in the greenhouse effect. Solid carbon dioxide (dry ice) is used in processes requiring large-scale refrigeration.

carbon monoxide

A colorless, odorless gas (CO), formed when carbon is oxidized in a limited supply of air. It is a poisonous constituent of automobile exhaust fumes, forming a stable compound with hemoglobin in the blood and preventing the transport of oxygen to the body tissues. In industry, carbon monoxide is used as a reducing agent in various metallurgical processes. It burns in air with a luminous blue flame to form carbon dioxide (CO_2).

carbonyl group

The group $=C=O$, which is found in many organic compounds such as ketones, aldehydes, carboxylic acids and amides. It also occurs in a number of inorganic carbonyl compounds in which carbon monoxide has coordinated with a metal atom or ion; for example, $Ni(CO)_4$.

carboxylic acid

Any organic acid containing the carboxyl group (–COOH) attached to another group,

CARBON CYCLE

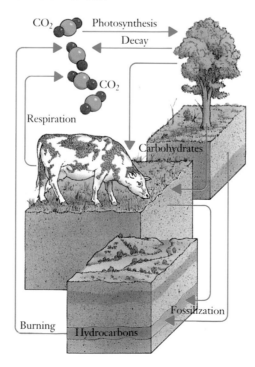

which can be hydrogen (giving methanoic acid) or a larger molecule (up to 24 carbon atoms). The smaller carboxylic acids form a homologous series, with all the names ending in the suffix *-oic* (for example, methanoic acid, HCOOH; ethanoic acid, CH_3COOH; and propanoic acid, C_2H_5COOH). Larger carboxylic acids may be found as esters in fats, often with glycerol, and are known as fatty acids.

casting

The operation of forming pourable mixes, usually of nonreinforced resins, into preshaped molds in which the material is polymerized with or without heat. Casting resins include certain acrylic, polyester, epoxy and polyurethane resins. This method is commonly used to encapsulate electrical components. Metals may also be cast.

catalyst

Any substance that makes possible or alters the rate of a chemical reaction but remains unchanged at the end of the reaction. Homogeneous catalysts are those that have the same phase as the reactants – in other words, catalysts and reactants that are all either liquids, solids or gases. Enzymes are homogeneous catalysts found in many biochemical reactions. Heterogeneous catalysts have a different phase from the reactions (for example, many of the metals used in industrial gas reactions). The catalyst works by lowering the **activation energy** of the chemical reaction by providing an alternative reaction route, thus increasing the rate at which the reaction reaches equilibrium. Most catalysts are used to speed up reactions and are highly specific in the reactions they catalyze. *See also* **catalytic converter**.

catalytic converter

The common name for a device fitted to the exhaust systems of many automobiles to reduce the amount of toxic emissions. A mixture of catalysts, usually based on platinum, thinly coated on a metal or ceramic honeycomb (to maximize the available surface area), is used to convert harmful substances in the passing exhaust gas into less harmful ones: for example, oxidation catalysts assist the conversion of hydrocarbons in unburned fuel and carbon monoxide into carbon dioxide and water. Three-way catalysts convert oxides of nitrogen back into nitrogen. Catalysts are "poisoned" by lead and sulfur compounds, and these must not be present in the fuel used by automobiles fitted with catalytic converters.

cathode

The negative electrode of an electrolytic cell, toward which positive particles (cations), usually in solution, are attracted (*see* **electrolysis**). A cathode is given its negative charge by connecting it to the negative side of an external electrical supply. This is in contrast to the negative electrode of an electrical (battery) cell, which acquires its charge in the course of a spontaneous chemical reaction taking place within the cell.

cation

A positively-charged ion. In an electrolytic cell, cations in the electrolyte are attracted to the negative electrode. *See* **cathode**.

cell, electrolytic

A device (cell) in which **electrolysis** is conducted.

cellulose

A complex carbohydrate (polysaccharide) made up of long unbranched chains of glucose units. Cellulose is a major component of the cell walls of all higher plants, and some algae and fungi, in which it provides rigidity to the cell wall. It is the most abundant substance found in the plant kingdom. Extracted cellulose has numerous uses in industry – for example, as a source of textiles (artificial silk, cotton, etc.) and plastics (cellophane and celluloid).

chain reaction

A succession of reactions, usually involving **free radicals**, in which the products of one stage are the reactants of the next. A chain

CHAIN REACTION

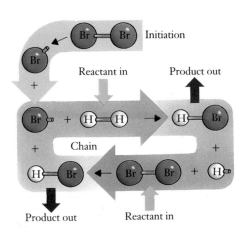

reaction is characterized by the continuous generation of reactive substances. It comprises three separate stages: **initiation** (the initial generation of reactive species), **propagation** (reactions involving reactive species that generate similar or different reactive species) and **termination** (reactions involving the reactive species but that produce only stable, nonreactive substances). Chain reactions may occur slowly (for example, the oxidation of edible oils) or accelerate as the number of reactive species increases, ultimately resulting in explosion. *See* **exothermic reaction** and **photochemistry**.

chemical energy

The energy that is stored in atoms and molecules and released during a chemical reaction. *See* **energy**.

chemical equation

A standard method of using chemical symbols and formulas to indicate the reactants and the products of a chemical reaction. A chemical equation gives two fundamental pieces of information: **1** the reactants (on the left-hand side) and products (right-hand side); and **2** the reacting proportions (stoichiometry; that is, how many units of each reactant and product are involved). The equation must balance so that the total number of atoms of a particular element on the left-hand side is the same as the number of atoms of that element on the right-hand side. For example, in the following equation, the sum of reactants equals the products:
$$Na_2CO_3 + 2HCl = 2NaCl + CO_2 + H_2O$$
This equation states that one molecule of sodium carbonate combines with two molecules of hydrochloric acid to form two molecules of sodium chloride, one of carbon dioxide and one of water.

In an equation, double arrows indicate that the reaction is reversible; for example, in the formation of ammonia from hydrogen and nitrogen, the direction of the reactions depends on the temperature and pressure of the reactants.

CONNECTIONS

BONDS AND STRUCTURES 56

HEAT IN AND HEAT OUT 64

chemical formula

A representation of a molecule, radical or ion, that uses the chemical symbols of the component elements. An empirical formula indicates the simplest ratio of the elements in a compound, without indicating how many of them there are or how they are combined. A molecular formula gives the

number of each type of element present in one molecule. A structural formula shows the relative positions of the atoms and the bonds between them. For example, for ethanoic acid, the empirical formula is CH_2O, the molecular formula is $C_2H_4O_2$ and the structural formula is CH_3COOH.

chemical name

The name given to a chemical substance. Trivial names were given to substances before their chemistry was fully understood, traditional names show the chemical composition of substances and the more recent systematic names give some idea of the structure of a substance. For example, blue vitriol (trivial), cupric sulfate (traditional) and copper(II) sulfate (systematic) are all names given to the same substance.

chemotherapy

Any medical treatment with chemicals. Examples include the treatment of cancer with cytotoxic and other drugs, and the treatment of many infections by penicillin. The term was coined by the German bacteriologist Paul Ehrlich for the use of synthetic chemicals against infectious diseases, beginning with his development of arsphenamine, the first effective drug against syphilis, in 1909–10.

CONNECTIONS

MEDICAL DRUGS 114

NATURAL DRUGS 116

TESTING FOR DRUGS 118

chirality

The property exhibited by a substance in which the molecules can exist in left- and right-handed structural forms. Most organic chiral molecules can be described in terms of chiral centers, in which an atom (usually carbon) has four different substituents. *See* **optical activity**.

chloride

The negative ion (Cl^-) formed when hydrogen chloride dissolves in water; and any salt containing this ion, commonly formed by the action of hydrochloric acid (HCl) on various metals or by direct combination of a metal and chlorine. Metals form ionic chlorides; sodium chloride (Na^+Cl^-) is common table salt. Nonmetals form covalent chlorides, such as tetrachloromethane (carbon tetrachloride, CCl_4).

chlorine

A greenish-yellow, gaseous, nonmetallic element with a pungent odor. Its symbol is Cl, its atomic number 17 and its relative atomic mass 35.453. Chlorine is a **halogen**. It is very reactive and is widely distributed in nature in combination with the alkali metals as chlorates or chlorides. In its elementary form the gas is a **diatomic molecule** (Cl_2). Chlorine is prepared industrially by the electrolysis of concentrated brine. It is used in making bleaches, in sterilizing water for drinking and for swimming pools, and in the manufacture of chloro-organic compounds such as chlorinated solvents, chlorofluorocarbons and polyvinyl chloride.

chlorofluorocarbons (CFCs)

A group of synthetic odorless, nontoxic, nonflammable and chemically inert chemicals that contain chlorine, fluorine, carbon and sometimes hydrogen. CFCs have been used as propellants in aerosol cans, as refrigerants in refrigerators and air conditioners, and in the manufacture of foam packaging. When CFCs are released into the atmosphere, they drift up slowly into the stratosphere. Under the influence of ultraviolet radiation from the Sun, they break down into chlorine atoms that destroy the ozone layer and allow harmful radiation from the Sun to reach the Earth's surface.

> **CONNECTIONS**
>
> CHEMISTRY OF LIFE **108**
>
> PHOTOSYNTHESIS **130**

chlorophyll

A green pigment that is present in most plants and is responsible for the absorption of light energy during photosynthesis. The pigment absorbs the red and blue-violet parts of sunlight but reflects the green, giving plants their characteristic color. Chlorophyll is similar in structure to **hemoglobin**, but with magnesium instead of iron as the reactive part of the molecule.

chromatography

An important technique for analyzing or separating mixtures of gases, liquids or dissolved substances. This is done by passing the mixture (mobile phase) through another substance (stationary phase), usually a liquid or solid. The different components of the mixture are absorbed or impeded to different extents, and hence separate. In paper chromatography, the mixture separates because the components have differing solubilities in the solvent flowing through the paper and in the chemically bound water of the paper. In thin-layer chromatography, a wafer-thin layer of adsorbent medium on a glass plate replaces the filter paper. The mix-

ture separates because of the differing solubilities of the components in the solvent flowing up the solid layer and their differing tendencies to stick to the solid (adsorption). The same principles apply in column chromatography. In **gas–liquid chromatography**, a gaseous mixture is passed into a long, coiled tube (heated in an oven) filled with an inert powder coated in a liquid. A carrier gas flows through the tube. As the mixture proceeds along the tube it separates as the components dissolve in the liquid to differing extents or stay as a gas. A detector locates the different components as they emerge from the tube. Analytical chromatography uses very small quantities, often millionths of a gram or less, to identify and quantify components of a mixture. Examples are the determination of the identities and amounts of amino acids in a protein, and the determination of the alcohol content of breath, blood and urine samples. Preparative chromatography is used for the large-scale purification and collection of one or more of the constituents; for example, the recovery of protein from effluent wastes.

colloid

A state of matter composed of extremely small particles of one material (the dispersed phase) evenly and stably distributed in another material (the continuous phase). The size of the dispersed particles is less than that of particles in suspension but greater than that of molecules in true solution. Colloids involving gases include aerosols (dispersions of liquid or solid particles in a gas, as in fog or smoke) and **foams** (dispersions of gases in liquids). Those involving liquids include **emulsions** (in which both the dispersed and the continuous phases are liquids) and **sols** (solid particles dispersed in a liquid). Sols in which both phases contribute to a molecular three-dimensional network have a jellylike form and are known as **gels**.

colorimetry

The quantitative analysis of solutions by comparing the color produced by a reagent with those produced in a standard solution.

combustion

Burning, defined in chemical terms as the rapid combination of a substance with oxygen, accompanied by the evolution of heat and usually light. A slow-burning candle flame and the explosion of a mixture of gasolene vapor and air are extreme examples of combustion. Incomplete combustion occurs when a substance has too little oxygen to burn completely. Most combustion processes are carried out to obtain the heat from combustion. However, some combustion

processes are used to produce specific products: for example, sulfur is burned in air to produce the sulfur dioxide required to produce sulfuric acid in the contact process.

> **CONNECTIONS**
>
> HEAT IN AND HEAT OUT **64**
>
> COMBUSTION AND FUELS **66**
>
> HYDROCARBON CHAINS **84**
>
> CARBON-HYDROGEN COMPOUNDS **86**

composite material

Any purpose-designed material created by combining other materials with complementary properties in a single composite. Most composites consist of a continuous matrix in which discrete elements (for example, aramid fibers) are dispersed. Examples are carbon fiber and glass-reinforced plastics. Composites may provide extra strength or wear resistance.

composition, chemical

The proportion by weight of each element in a chemical compound. The law of definite proportions states that pure compounds have a fixed and invariable composition. The law of multiple proportions states that, if two elements form more than one compound, the various weights of one which combine with a given weight of the other are in small, whole-number ratios.

compound

Any substance made up of two or more elements bonded together, such that they cannot be separated by physical means. The formation of a compound requires a **chemical reaction**. Compounds are held together by ionic or covalent bonds. *See also* **molecule**.

> **CONNECTIONS**
>
> MIXTURES AND COMPOUNDS **52**
>
> TYPES OF BONDS **54**

compression molding

A method for producing shaped material by placing the material in a heated, hardened, polished steel molding vessel and forcing it down by means of a plunger at high pressure (between 20 and 35 MN/m^2).

concentration

The quantity of a substance (solute) present in a unit volume of a solution. It is commonly measured in moles per cubic decimeter and moles per liter. The term also refers to the process of increasing the concentration of a solution by removing some of the

CONDENSATION

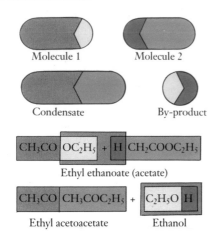

Molecule 1 — Molecule 2

Condensate — By-product

$$CH_3CO \mid OC_2H_5 + H \mid CH_2COOC_2H_5$$

Ethyl ethanoate (acetate)

$$CH_3CO \mid CH_3COC_2H_5 + C_2H_5O \mid H$$

Ethyl acetoacetate — Ethanol

substance (solvent) in which the solute is dissolved. In a **concentrated solution**, the solute is present in large quantities.

condensation

1 In physical chemistry, the conversion of a vapor to a liquid as it loses heat. This is frequently achieved by letting the vapor come into contact with a cold surface and is an essential step in distillation processes. **2** In organic chemistry, any reaction in which two organic compounds combine to form a larger molecule, accompanied by the removal of a smaller molecule (usually water). This reaction is also known as an addition–elimination reaction. Condensation polymerization is a polymerization reaction in which one or more monomers, with more than one reactive functional group, combine to form a polymer. Polyamides (such as nylon) and polyesters (such as Dacron) are made by condensation **polymerization.**

coordinate bond
See **bond.**

copolymer

Any polymer produced by the polymerization of more than one kind of monomer. *See* **polymerization.**

covalent bond

The bond produced when two atoms share one or more pairs of electrons (usually each atom contributes an electron). The bond is often represented by a single line drawn between the two atoms. Double bonds – found, for example, in the **alkenes** – are formed when two atoms share two pairs of electrons (the atoms usually contribute a pair each); triple bonds, seen in the **alkynes**, are formed when atoms share three pairs of electrons. Such bonds are represented by a double or triple line, respectively, between the

atoms concerned. Covalent compounds have low melting and boiling points; are poor conductors of electricity; and are usually insoluble in water and soluble in organic solvents. *See* **bond** and **carbon–carbon bond.**

cracking

The process of breaking down chemical compounds by heat. The term cracking is often used to refer to petroleum refining, where large alkane molecules are broken down by heat into smaller alkanes and alkene molecules to produce branched-chain hydrocarbons suitable, for example, for gasoline. The reaction is carried out at a high temperature and high pressure, and often in the presence of a **catalyst** (catalytic cracking).

CONNECTIONS

CARBON-HYDROGEN COMPOUNDS 86
SYNTHETIC POLYMERS 98

crosslink

The short lateral linking of two or more long-chain molecules in a polymer. Crosslinking gives the polymer a higher melting point and increases its rigidity. A common example of a crosslinked polymer is vulcanized rubber. *See also* **elastomer.**

crystal

Any solid substance with an orderly three-dimensional arrangement of its atoms or molecules, thereby creating an external surface of clearly defined smooth faces. Crystals have certain physical characteristics, such as the angles between the faces and the refractive index. Crystals fall into a number of crystal systems or groups, classified on the basis of the relationship of a number of imaginary axes that intersect at the center of any perfect, undistorted crystal. A mineral

CRYSTAL

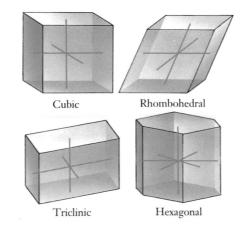

Cubic — Rhombohedral

Triclinic — Hexagonal

can often be identified by the shape of its crystals. Common examples of crystal structure can be found in table salt (cubic) and quartz (rhombohedral or trigonal).

cyanide

The ion (CN^-) derived from hydrogen cyanide (HCN), and any salt containing this ion (produced when hydrogen cyanide is neutralized by alkalis), such as potassium cyanide (KCN). The principal cyanides are those of potassium, sodium, calcium, mercury, gold and copper. *See also* **nitrile.**

cyclic compound

Any of a group of organic compounds that have rings of atoms in their molecules. They may be alicyclic (such as cyclopentane), aromatic (benzene) or heterocyclic (pyridine). Alicyclic compounds (aliphatic cyclic) have localized bonding: all the electrons are confined to their own particular bonds, in contrast to **aromatic compounds**, in which certain electrons have free movement between different bonds in the ring. Alicyclic compounds have chemical properties similar to their straight-chain counterparts; aromatic compounds, because of their special structure, undergo entirely different chemical reactions. Heterocyclic compounds have a ring of carbon atoms with one or more carbons replaced by another element, usually nitrogen, oxygen or sulfur. They may be aliphatic or aromatic.

dative covalent bond
See **bond.**

deliquescence

The absorption of so much moisture from the air by a hygroscopic (water-absorbing) solid so that the solid ultimately dissolves to form a solution. Deliquescent substances such as calcium chloride, potassium hydroxide and sodium hydroxide make very good drying agents and are used in the bottom chambers of desiccators (laboratory equipment for drying substances).

desulfurization

The process of removing sulfur and sulfur-containing compounds from emission gases to avoid the formation of the sulfur-based acids that contribute to acid rain.

detergent

A substance added to water to improve its cleaning ability. Common detergents are made from fats and sulfuric acid, and their long-chain molecules have a salt group at one end attached to a long hydrocarbon "tail" – a structure similar to that of soap molecules. In order to remove dirt, generally attached to materials by means of oil or grease, the hydrocarbon "tails" (soluble in oil or grease) penetrate the oil or grease drops, while the "heads" (soluble in water but insoluble in grease) remain in the water and, being salts, become ionized. Consequently the oil drops become negatively charged and tend to repel one another; thus they remain in suspension and are washed away with the dirt. Detergents have the advantage over soap in that they do not produce scum by forming insoluble salts with the calcium and magnesium ions present in hard water. Phosphates in some detergents can cause the excessive enrichment of rivers and lakes.

CONNECTIONS

BONDS AND STRUCTURES **56**

SOAPS AND DETERGENTS **80**

---

developer

In **photography**, a reducing agent based on a **benzene ring** structure substituted by two or three hydroxy or amino groups. The developer reduces the grains of silver halides (usually bromides) in the **latent image** to produce a **negative** image. Hydroquinone has been a historically important constituent of developers, which may also contain wetting agents and bactericides.

diabetes

The disease in which a disorder of the islets of Langerhans in the pancreas prevents the body producing enough of the hormone insulin, so that sugars cannot be properly utilized. Treatment is by dietary control, chemotherapy or injected insulin.

digestion

The process in which food consumed by an animal is broken down. This occurs both physically and by a series of chemical reactions involving enzymes to make the nutrients available for absorption and cell metabolism.

dissociation

The process in which a single compound is split into two or more smaller products, which may be capable of recombining to form the reactant. Where dissociation is incomplete (not all the compound's molecules dissociate), a chemical equilibrium exists between the compound and its dissociation products. Dissociation may be brought about by heat (thermal dissociation) or by dissolving the substance in water to form **ions** (ionization). Dissociation accounts in part for the phenomenon of electrolytic conductivity.

DYE

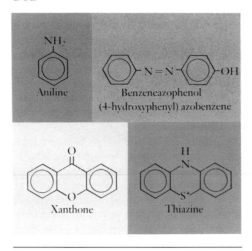

Aniline

Benzeneazophenol
(4-hydroxyphenyl) azobenzene

Xanthone

Thiazine

---

distillation

A technique used to purify liquids or to separate mixtures of liquids possessing different boiling points. Simple distillation is used in the purification of liquids (or the separation of substances in solution from their solvents); for example, in the production of pure water from a salt solution. The solution is boiled and the vapors of the solvent rise into a separate piece of apparatus (the condenser) where they are cooled and condensed. The liquid produced (the distillate) is the pure solvent; the non-volatile solutes (now in solid form) remain in the distillation vessel to be discarded as impurities or recovered as required. Mixtures of liquids (such as petroleum or aqueous ethanol) are separated by **fractional distillation**.

CONNECTIONS

CARBON-HYDROGEN COMPOUNDS **86**

CARBON, HYDROGEN AND OXYGEN **88**

---

double bond

Two covalent bonds between adjacent atoms, as in the alkenes ($-C=C-$) and ketones ($-C=O$). *See* **covalent bond**.

drug

Any of a range of naturally produced or synthetic chemicals introduced into the body to enhance or suppress a biological function. Most drugs now in use are medicines (pharmaceuticals), used to prevent or treat diseases, or to relieve their symptoms. Many drugs are similar to chemicals that are found naturally in the body. They include hormones, antibiotics, cytotoxic drugs (drugs to kill diseased or cancerous cells), immunosuppressives, sedatives and pain relievers (analgesics). *See* **chemotherapy** and **pharmacology**.

dye

Any substance that, when applied in solution, imparts a color to a substrate, such as hair or textiles. Direct dyes combine with the material of the fabric to yield a colored compound; indirect dyes require the presence of another substance (*see* **mordant**), with which the fabric must first be treated; vat dyes are colorless soluble substances that on exposure to air yield an insoluble colored compound. Naturally occurring dyes include indigo, madder (alizarin), logwood and cochineal; however, industrial dyes (introduced in the 19th century) are usually synthetic. Industrial dyes include azo dyestuffs and aniline dyes.

elastomer

Any natural or synthetic rubber or material with rubbery properties that stretches easily and then quickly returns to its original length when released. Examples include polychloroprene and butadiene copolymers. The convoluted molecular chains making up these materials are uncoiled by a stretching force, but return to their original position when released because there are relatively few **crosslinks** between the chains.

electrode

Any terminal by which an electric current passes in or out of a conducting substance; for example, the anode or cathode in an electrolytic cell (*see* **electrolysis**).

electrolysis

The process of passing an electric current through a solution or molten salt (the **electrolyte**), resulting in the migration of ions to the electrodes: positive ions (**cations**) to the negative electrode (**cathode**) and negative ions (anions) to the positive electrode (**anode**). During electrolysis, the ions react with the electrodes, either receiving or giving up electrons. The resultant atoms may be liberated as a gas or deposited as a solid on the electrode. For instance, when acidified water is electrolyzed, hydrogen ions (H^+) at the cathode receive electrons to form hydrogen gas; hydroxide ions (OH^-) at the anode give up electrons to form oxygen gas and water.

A common application of electrolysis is **electroplating**, in which a solution of a salt, such as silver nitrate ($AgNO_3$), is used. The object to be plated acts as the negative electrode, thus attracting silver ions (Ag^+). Electrolysis is used in many industrial processes, such as coating metals for vehicles and ships, and refining bauxite into aluminum; it also forms the basis of a number of electrochemical analytical techniques, such as polarography.

electrolyte

A solution or molten substance in which an electric current is made to flow as a result of the presence of positive or negative ions (*see* **electrolysis**). The term "electrolyte" is frequently used to refer to a substance that, when dissolved in a specified solvent (usually water), produces an electrically conducting medium. Liquid metals, in which the conduction is by free electrons, are not usually considered to be electrolytes.

electron

A stable, negatively charged elementary particle that is a component of all atoms. The electrons in each atom surround the nucleus in groupings called shells; in a neutral atom the number of electrons is equal to the number of protons in the nucleus. This electron structure is responsible for the chemical properties of the atom, and for the ways in which it combines with other atoms to form molecules. Electrons are also the basic particles of electricity. Each carries a charge of 1.602192×10^{-19} coulomb. A beam of free electrons undergoes diffraction and produces interference patterns in the same way as electromagnetic waves such as light; similarly, it can be focused using magnetic lenses, a property that allows electron microscopes.

CONNECTIONS

THE ELEMENTS **50**

MIXTURES AND COMPOUNDS **52**

TYPES OF BONDS **54**

BONDS AND STRUCTURES **56**

NAMES AND FORMULAS **60**

---

electronic balance

A very accurate type of **balance**. Weights are determined by a scanner measuring the displacement of a measuring pan and generating an equivalent electrical signal that can be read on a digital display.

electroplating

A method of plating one metal with another by **electrolysis**. A current is passed through a bath containing a solution of a salt of the plating metal, the object to be plated being the **cathode**. The **anode** is either an inert substance or the plating metal; for example, zinc, nickel, chromium, cadmium, copper, silver and gold.

element

Any substance that cannot be split chemically into simpler substances. The atoms of a particular element all have the same number of protons in their nuclei (their atomic number). Elements are classified in the **Periodic** Table of the elements. Of the 106 known elements, 95 are known to occur naturally (those with atomic numbers from 1 to 95). Those with atomic numbers from 96 to 106 are synthesized, produced in particle accelerators. Eighty-one of the elements are stable; all the others, which include atomic numbers 43, 61 and from 84 up, are radioactive. Elements are classified as metals, non-metals or metalloids (weakly metallic elements) depending on a combination of their physical and chemical properties; about 75 percent are metallic. Some elements occur abundantly (oxygen, aluminum); others occur moderately or rarely (chromium, neon); some, in particular the radioactive ones, are found in minute amounts (neptunium, plutonium). Symbols are used to denote the elements; the symbol is usually the first letter or letters of the English or Latin name (for example, C for carbon; Ca for calcium; Fe for iron, ferrum). The symbol represents one atom of the element. According to current theories, hydrogen and helium were produced in the Big Bang at the beginning of the Universe. Of the other elements, those up to atomic number 26 (iron) are made by nuclear fusion within the stars. The more massive elements, such as lead and uranium, are believed to be produced when an old star explodes; as its center collapses, the gravitational energy squashes nuclei together to make new elements.

emulsifier

Any substance used to stabilize emulsions of water dispersed in an oil or an oil dispersed in water. Egg yolk is a naturally occurring emulsifier. Commercial synthetic emulsifiers, such as detergents, have different affinities for oils and water at different sites in their molecules.

emulsion

A stable dispersion (*see* **colloid**) of one liquid in another liquid; for example, oil and water emulsions are used in some cosmetic lotions, and milk is an emulsion of fats and water. *See* **emulsifier**.

CONNECTIONS

MIXTURES AND COMPOUNDS **52**

MAKING HYDROCARBONS **92**

PAINTS, PIGMENTS AND INKS **122**

---

emulsion paint

A paint made up of a pigmented emulsion, usually of a synthetic resin in water; water can be used to thin it. Resins that are used in emulsion paints include polyvinyl acetate and polyvinyl chloride.

enantiomer

One of two mirror image forms of a substance in which the molecules are asymmetric. A classic example is that of the crystals of sodium ammonium tartrates, which the French scientist Louis Pasteur showed to exist in mirror image forms. Enantiomers are also known as optical isomers. *See* **optical activity**.

endothermic reaction

Any physical or chemical change in which energy is absorbed by the reactants from the surroundings. The energy absorbed is represented by the symbol ΔH. The dissolving of sodium chloride in water is a physical endothermic change. The hardening of egg white (albumen) in cooking an egg is an example of a chemical endothermic reaction. *See also* **exothermic reaction**.

energy

The capacity for doing work. Potential energy is energy deriving from position: a stretched spring has elastic potential energy, and an object raised to a height above the Earth's surface, or the water in an elevated reservoir, has gravitational potential energy. A lump of coal and a tank of gasoline, together with the oxygen needed for their combustion, have **chemical energy**. Other kinds of energy include light, sound and electrical and nuclear energy. Moving bodies possess kinetic energy. Energy can be converted from one form to another, but the total quantity stays the same (in accordance with the conservation of energy principle).

enzyme

A **protein** molecule that acts as a **catalyst** to speed up biochemical reactions in the conversion of one molecule (substrate) into another. Enzymes are large, complex proteins and are highly specific, each reaction requiring its own particular enzyme. The enzyme fits into a "slot" (active site) in the substrate molecule, forming an enzyme-substrate complex that lasts until the substrate is altered or split, after which the enzyme can fall away. The substrate may therefore be compared to a lock and the enzyme to the key required to open it. Enzymes have many medical and industrial uses, from components of soap and detergents to drug production, and as research tools in molecular biology.

CONNECTIONS

CHEMISTRY OF LIFE **108**

VITAL RAW MATERIALS **110**

LIVING CHEMISTRY **112**

---

equation
See **chemical equation**.

equilibrium (chemical)
A state in which the energy of a particular chemical system is distributed in the most probable way. Chemical equilibrium is reached in a reversible reaction when the reaction and its reverse are proceeding at equal rates, so that the system has no further tendency to change.

equivalent weight
That quantity of a substance that reacts chemically with a given amount of a standard (for example, with one gram of hydrogen). It represents the "combining power" of a substance. For a compound it depends on a particular reaction, whereas for an element it is given by the **relative atomic mass** divided by the **valence**.

ester
An organic compound formed by the reaction between an **alcohol** and an **acid**, with the elimination of water. Unlike salts, esters are covalent compounds. Esters formed from carboxylic acid groups have the general formula R.COO.R. Esters containing simple hydrocarbon groups are fragrant, volatile substances that are frequently used as flavorings in the food industry.

ethanal
Also known as acetaldehyde (CH_3CHO), one of the chief members of the group of organic compounds known as aldehydes. It is a colorless flammable liquid that boils at 20.8°C. Ethanal is formed by the oxidation of ethanol or ethene and is used in the manufacture of many other organic chemical compounds.

ethane
A colorless, odorless and flammable gaseous **hydrocarbon** (CH_3CH_3). It is the second member of the **alkane** series of hydrocarbons (paraffins).

ethanoate
Also known as **acetate** (CH_3COO^-), a negative ion derived from ethanoic acid or any salt containing this ion. In textiles, acetate rayon is a synthetic fabric made from modified cellulose treated with ethanoic acid; in photography, cellulose ethanoate is used to make nonflammable acetate film.

ethanoic acid
Also known as **acetic acid** (CH_3COOH), one of the simplest **fatty acids**. In its pure state it is a colorless liquid with a characteristic pungent odor. It solidifies to an icelike mass of crystals at 16.7°C and is often called glacial ethanoic acid. Vinegar contains five percent or more ethanoic acid, produced by fermentation. Cellulose (derived from wood or other sources) may be treated with ethanoic acid to produce a cellulose ethanoate (acetate) solution, which can be used to make plastic items by molding or extruded to form synthetic textile fibers.

ethanol
Also known as ethyl **alcohol** (C_2H_5OH), ethanol is the alcohol found in alcoholic drinks. When pure, it is a colorless liquid with a pleasant odor, and can be mixed with water or ether; it burns in air with a pale blue flame. Ethanol vapor forms an explosive mixture with air and may be used in high-compression internal combustion engines. Ethanol is produced naturally by the fermentation of carbohydrates by yeast cells. It can be made industrially by absorption of ethene and subsequent reaction with water, or by reducing ethanal in the presence of a catalyst. It is widely used as a solvent.

ethene
Also known as ethylene (C_2H_4), a colorless, flammable gas, the first member of the alkene series of hydrocarbons. It is the most widely used synthetic organic chemical and is employed to produce the plastics polyethene (polyethylene), polychloroethene and polyvinyl chloride. Ethene is obtained from natural gas or coal gas, or by the dehydration of ethanol.

EXTRUSION

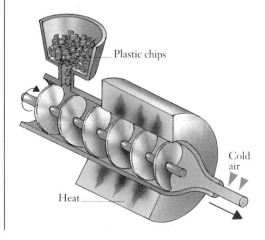

Plastic chips

Cold air

Heat

ethylene
See **ethene**.

ethyl group
The organic group $-C_2H_5$.

ethyne
A traditional name for **acetylene** (CHCH), a colorless flammable gas. It is the simplest member of the **alkyne** series of **hydrocarbons** and is widely used in the manufacture of synthetic rubbers. Ethyne was discovered by the British scientist Edmund Davy in 1836 and was used in early gas lamps, where it was produced by the reaction between water and calcium carbide. Its combustion provides more heat, relatively, than almost any other fuel known; it is widely used in oxyacetylene welding and cutting.

evaporation
The process by which a liquid turns to a vapor without its temperature reaching boiling point. A liquid left open to the air eventually evaporates because a proportion of its molecules have enough kinetic energy to escape through the intermolecular forces of attraction at the liquid surface into the atmosphere. The temperature of the liquid tends to fall because the evaporating molecules remove energy.

exhaust emission
Emissions produced by automobiles burning traditional fossil fuel based on gasoline in an internal combustion engine. They include toxic gases such as **carbon monoxide**, **nitrogen oxides** and **hydrocarbons**, together with other components of gasoline, such as lead. *See also* **catalytic converter**.

exothermic reaction
Any physical or chemical change, such as combustion, in which energy is released by the reactants and passed into the surroundings. An explosion is a very rapid exothermic reaction. *See also* **endothermic reaction**.

extender
Any inert substance added to a commercial product such as paint or washing powder, to dilute it for economy or to alter its physical properties; for example, adjusting the viscosity of an adhesive.

extrusion
A process used in the production of artificial textiles. A viscous solution of the polymer being used, or the molten polymer, is forced through fine holes in a metal dye (**spinneret**) to produce fibers. Extrusion is also used for manufacturing plastics and metal products such as tubes.

fastness

The ability of a color dye to remain unchanged when exposed to a specific agent, such as light, wear, chemical action or damp.

fat

A mixture of lipids (chiefly **triglycerides**) that is solid at room temperature. Lipid mixtures that are liquid at room temperature are called **oils**. The higher the proportion of **saturated** fatty acids in a mixture, the harder the fat. Fats are essential constituents of food for many animals, with a calorific value twice that of carbohydrates.

fatty acid

Any organic compound consisting of a **hydrocarbon** chain, up to 24 carbon atoms long, with a carboxyl group (–COOH) at one end. The covalent bonds between the carbon atoms may be single or double; where a double bond occurs, the carbon atoms carry one instead of two hydrogen atoms. Chains with only single bonds have all the hydrogen they can carry, so they are said to be **saturated** with hydrogen. Chains with one or more double bonds are said to be **unsaturated** (or polyunsaturated).

Saturated fatty acids include palmitic and stearic acids; unsaturated fatty acids include oleic (one double bond), linoleic (two double bonds) and linolenic (three double bonds). Unsaturated fatty acids are generally liquid at room temperature, whereas saturated fatty acids have a higher melting point. Fatty acids are generally found combined with glycerol in lipids such as triglycerides.

FATTY ACID

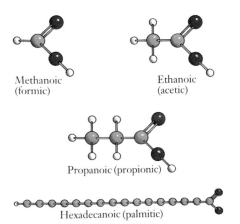

Methanoic (formic)

Ethanoic (acetic)

Propanoic (propionic)

Hexadecanoic (palmitic)

fertilizer

A substance used to rectify the deficiencies of poor or depleted soil. A fertilizer contains some or all of a range of about 20 chemical elements necessary for healthy plant growth. Fertilizers may be organic – farmyard manure, composts, bonemeal – or inorganic, in the form of compounds, mainly of nitrogen, phosphate and potash. Externally applied fertilizers usually contain more nutrients than plants require, and may pollute lakes and rivers when they drain off. In view of environmental concerns, research is now focused on the actual modification of crop plants themselves – for example, in trying to establish a symbiotic relationship between nitrogen-fixing bacteria found in the root nodules of legumes and crop plants such as wheat, so as to allow nitrogen to be assimilated directly from the atmosphere.

fiber

Synthetic fibers, used in textiles, are produced by synthesizing a suitable linear polymer and extruding the molten polymer to produce a fiber of the required cross-section. *See* **spinneret**.

fire extinguisher

Any device for putting out a fire by removing either heat, oxygen or fuel – the three conditions necessary for the fire to continue. Basic fire extinguishers propel water onto the fire. Water extinguishers cannot be used on electrical fires, due to the danger of electrocution, or on burning oil, because the oil will float on the water and spread the blaze. Many domestic extinguishers contain pressurized liquid carbon dioxide. Carbon dioxide is released as a gas that blankets the burning material and prevents oxygen from reaching it. Dry extinguishers spray powder, which then releases carbon dioxide gas. Wet extinguishers are often of the soda-acid type; when activated, sulfuric acid mixes with sodium hydrogen carbonate to produce carbon dioxide. Some extinguishers contain halons (hydrocarbons with one or more hydrogen atoms substituted by a halogen such as chlorine, bromine or fluorine). These are very effective at smothering fires, but cause damage to the ozone layer.

flammability

The liability of a material to catch fire. Many synthetic materials are highly flammable and need to be treated with an anti-inflammatory coating prior to commercial application.

flash point

The minimum temperature at which a liquid or volatile solid heated under standard conditions gives off sufficient vapor to ignite on the application of a small flame. The fire point of a material is the temperature at which full combustion occurs. For safe storage of materials such as fuel or oil, conditions must be well below the flash and fire points to minimize the risk of fire. *See also* **spontaneous combustion**.

fluidized-bed reactor

A reaction vessel in which the particles of the active material are separated from each other by a high velocity fluid (gas or liquid) passing upwards through a bed of the solid material. Such beds are commonly used in the chemical industry to allow intimate contact between a solid and a gas, high rates of heat transfer, and uniform temperatures within the bed.

fluorine

A pale yellow, gaseous, nonmetallic element. Its symbol is F, its atomic number 9, and its relative atomic mass 19. Fluorine is the first member of Group VIIA of the Periodic Table (the **halogens**), and is pungent, poisonous and reactive, uniting directly with nearly all the elements. It occurs naturally as the minerals fluorite (CaF_2) and cryolite (Na_3AlF_6). Hydrogen fluoride is used in etching glass; freons, organic compounds which all contain fluorine and chlorine or bromine, are widely used as refrigerants.

foam

A dispersion of bubbles in a liquid. Solid foams such as expanded polystyrene or foam rubber are made by first foaming the liquid and then allowing it to set. Foams can be stabilized by **foam stabilizers**. They are sometimes known as froths.

foam stabilizer

Agents such as oils, soaps or emulsifying agents that help to stabilize foams by the production of a binding network in the walls of bubbles. They work by increasing the spreading or wetting properties of a liquid by reducing its surface tension. Foam stabilizers are also known as surfactants (surface active agents).

formaldehyde

See **methanal**.

formic acid

See **methanoic acid**.

formula
See **chemical formula**.

fractional distillation
The process used to split complex mixtures (such as crude oil) into their components, usually by repeated heating, boiling and condensation (distillation). Good separation can be achieved by using a long fractionating column filled with glass beads attached to the distillation vessel. The rising vapor flows over the descending liquid to create a steady state in which there is a decreasing temperature gradient in the column. Various fractions (a group of similar compounds, the boiling points of which fall within a particular range) can then be drawn off at various points on the column.

CONNECTIONS

HYDROCARBON **84**

CARBON-HYDROGEN COMPOUNDS **86**

CARBON, HYDROGEN AND OXYGEN **88**

fractionation
See **fractional distillation**.

Frasch process
The process used to extract underground deposits of sulfur, using a tube consisting of three concentric pipes. Superheated steam at 165°C is piped down the outer pipe into the sulfur deposit and melts it. Compressed air is then pumped down the inner pipe to force a frothy mixture of molten sulfur and water to the surface through the middle pipe. The steam in the outer casing keeps the sulfur molten in the pipe until it reaches the surface. The process, which results in very pure sulfur, was developed in the United States in 1891 by the German-born scientist Herman Frasch.

FRACTIONAL DISTILLATION

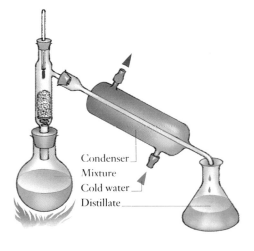

Condenser
Mixture
Cold water
Distillate

free radical
An atom or molecule that has an unpaired electron and is therefore highly reactive. Most free radicals are very short-lived. If free radicals are produced in living organisms they can be very damaging. They are often produced by high temperatures and are found in flames and explosions. A simple free radical is the methyl radical $.CH_3$, which is produced by the splitting of the covalent carbon–carbon bond in ethane. The action of ultraviolet radiation from the Sun splits **chlorofluorocarbon** molecules in the upper atmosphere into free radicals, which then break down the ozone layer. *See* **carbon–carbon bond** and **covalent bond**.

freezing point
The temperature for any given liquid at which any further removal of heat will convert the liquid into the solid state. The temperature remains at this point until all the liquid has solidified. It is invariable under similar conditions of pressure; for example, the freezing point of pure water under standard atmospheric pressure is 0°C.

Friedel–Crafts reaction
A chemical reaction used to synthesize alkyl-substituted benzene hydrocarbons and aromatic ketones. It uses the action of halogenoalkanes or acyl halides on aromatic hydrocarbons, catalyzed by anhydrous aluminum chloride. The reaction is named after the French chemist Charles Friedel and the United States chemist James M. Crafts.

fuel
Any substance that is oxidized or otherwise changed by being burned in a heat engine or furnace to release useful energy or heat. Fuels cover the entire range of materials that burn (combustibles). Wood, vegetable oils and animal products have largely been replaced since the 18th century by fossil fuels such as oil, gas and coal. The limited supply of fossil fuels and the expense of extracting them has led to the development of nuclear fuels and other, renewable sources of energy to produce electricity.

CONNECTIONS

HEAT IN AND HEAT OUT **64**

COMBUSTION AND FUEL **66**

CARBON-HYDROGEN COMPOUNDS **86**

MAKING HYDROCARBONS **92**

gas
A form of matter, such as air, in which the molecules move randomly in otherwise empty space, filling any size or shape of container into which the gas is put. Gases can be liquefied by cooling, which lowers the speed of the molecules and enables attractive forces between them to bind them together.

gas–liquid chromatography
A very sensitive form of **chromatography** in which the mobile phase (the mixture to be analyzed) is a gas and the stationary phase is a liquid. Solid and liquid substances are vaporized before being introduced into the apparatus.

gasoline
A colorless and highly volatile mixture of hydrocarbons, each containing between five and eight carbon atoms, that is derived from petroleum. Gasoline is mainly used as a fuel for internal-combustion engines and for making other chemicals.

gel
A **colloid** in which the two phases combine to produce a solid or semisolid material.

glucose
Also known as dextrose or grape sugar ($C_6H_{12}O_6$), glucose is a naturally occurring sugar produced from other sugars and starches. It forms the "energy currency" of biochemical respiration reactions. Glucose is a monosaccharide (made up of a single sugar unit), unlike the more familiar sucrose (cane or beet sugar), which is a disaccharide (made up of two units: glucose and another sugar, fructose). Glucose is prepared in syrup form by the hydrolysis of cane sugar or starch, and may be purified to a crystalline powder.

CONNECTIONS

CARBON, HYDROGEN AND OXYGEN **88**

VITAL RAW MATERIALS **110**

LIVING CHEMISTRY **112**

glycerol
Also known as glycerine or propan–1,2,3–triol ($HOCH_2CH(OH)CH_2OH$), a thick, colorless, odorless, sweetish liquid. It is obtained from vegetable and animal oils and fats or by fermentation of glucose. Glycerol is used in the manufacture of high explosives, in antifreeze solutions, to keep fruits and tobacco moist, and in cosmetics.

glycogen
A polysaccharide of glucose made and retained in the liver as a carbohydrate store; for this reason it is sometimes called an animal starch. Glycogen provides energy to muscles often being converted back into glucose by the hormone insulin and metabolized.

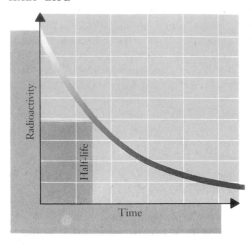

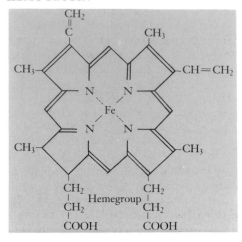

gravimetric analysis

A type of quantitative chemical analysis for determining the amount of a particular substance present in a sample by weighing it. This usually involves the conversion of the test substance into a compound of known molecular weight that can be easily isolated and purified.

greenhouse effect

The phenomenon in the Earth's atmosphere by which heat from the Sun, trapped by the Earth and re-emitted from its surface, is prevented from escaping by various gases in the atmosphere. The main greenhouse gases are water vapor, carbon dioxide, methane (CH_4) and **chlorofluorocarbons** (CFCs). Carbon dioxide and methane both occur naturally, and the amounts produced industrially, though significant to the greenhouse effect, form only a small proportion of the overall production of these gases. Carbon dioxide is mainly produced by the burning of fossil fuels and large tracts of forests (*see* **carbon cycle**). Methane is a major byproduct of several industrial processes.

Haber process

The industrial process for the production of ammonia (NH_3) by the reversible reaction of hydrogen with atmospheric nitrogen. A temperature of about 400°C is used with a catalyst of finely divided iron containing potassium and aluminum oxide promoters, at a pressure of about 250 atmospheres. About 10 percent of the reactants combine and the uncombined gases are recycled. The ammonia is separated either by being dissolved in water or by being cooled to liquid form. The process is very important for the fixation of nitrogen for synthetic fertilizers. It was originally developed by the German chemist Fritz Haber in 1908. (It is also known as the Haber–Bosch process.)

half-life

The time during which the strength of a radioactive isotope decays to half its original value. In theory, for any sample of the isotope, the decay process is never complete and there is always some residual radioactivity. For this reason, the half-life is measured rather than the total decay time. It may vary from millionths of a second to billions of years. Carbon-14, a naturally occurring isotope, has a half-life of 5730 years. The proportions of undecayed isotopes with known half-lives is used in radioisotope dating to date ancient organic material.

Hall–Héroult process

The industrial process used for extracting aluminum from bauxite by electrolysis. The bauxite is first purified by being dissolved in sodium hydroxide and filtering off the insoluble material. Aluminum hydroxide is precipitated and decomposed by heating to obtain pure Al_2O_3. This oxide is then mixed with cryolite to lower its melting point and the mixture electrolyzed using graphite electrodes. Molten aluminum collects at the bottom of the electrolytic cell and can be tapped off. The process is named for the United States chemist Charles Martin Hall and the French chemist Paul Héroult, who each developed it independently in 1886.

halogen

Any of a group of five nonmetallic elements with similar chemical bonding properties: fluorine, chlorine, bromine, iodine and astatine. They form a group (Group VIIA) in the **Periodic Table** of the elements, descending from fluorine (the most reactive) to astatine (the least reactive). The halogens combine directly with most metals to form salts, such as common salt (NaCl). Each has seven electrons in its valence shell, resulting in chemical similarities displayed by the group.

halogenation

A reaction in which a **halogen** atom is incorporated into a compound. Halogenation may take place by direct reaction with the halogen, through electrophilic substitution using aluminum chloride as a catalyst or using halogenating compounds, such as phosphorus halides, which react with hydroxyl groups.

helium

A colorless, odorless, gaseous, nonmetallic element, with the symbol He, the atomic number 2, and relative atomic mass 4.0026. It is grouped with the inert gases, is nonreactive and forms no compounds. It is the second most abundant element (after hydrogen) in the Universe, and has the lowest boiling (–268.9°C) and melting points (–272.2°C) of all the elements. The helium nucleus, consisting of two protons and two neutrons, is given off in alpha decay radioactivity. It is present in small quantities in the Earth's atmosphere from gases issuing from radioactive elements in the Earth's crust. Gaseous helium is lighter than air and nonflammable, and is therefore used for high-altitude balloons and airships.

hemoglobin

One of a group of globular proteins used by all vertebrates and some invertebrates for transporting oxygen and carbon dioxide around the body. In vertebrates it occurs in red blood cells (erythrocytes), and gives them their color. The hemoglobin molecule consists of four polypeptide chains (the globin protein) each bearing a heme group (an iron-containing molecule) which can reversibly bind oxygen and carbon dioxide.

homologous series

Any of a number of series of organic chemicals in which the molecular formulas, when arranged in ascending order, form an arithmetical progression. **Alkanes** (paraffins), **alkenes** (olefins) and **alkynes** (acetylenes) form such series in which members differ in mass by 14 atomic mass units. Thus the alkane homologous series begins with methane (CH_4), ethane (C_2H_6), propane (C_3H_8), butane (C_4H_{10}), and pentane (C_5H_{12}); each member of this series differs from the previous one by a CH_2 group (or 14 atomic mass units). The members of the group have chemical similarities, but their physical properties undergo a gradual change from one member to the next.

hormone

A product of an endocrine gland that is secreted in very small quantities into the bloodstream. Hormones bring about changes in the functions of various organs according to the body's requirements. The pituitary gland is a center for coordination of hormone secretion; the thyroid hormones determine the rate of general body chemistry; the adrenal hormones prepare the organism during stress; and the sex hormones, such as estrogen, govern reproductive functions.

hydrate

Any chemical compound that has discrete water molecules combined with it. The water is known as water of crystallization and the number of water molecules associated with one molecule of the compound is denoted in both its name and chemical formula. For example, $CuSO_4.5H_2O$ is copper(II) sulfate pentahydrate.

hydride

Any binary compound of hydrogen and another element. There are three types: covalent hydrides, ionic hydrides and metallic hydrides.

hydrocarbon

Any of a class of chemical compounds containing only hydrogen and carbon. Many different hydrocarbons exist, including aliphatic, alicyclic and aromatic hydrocarbons. Aliphatic hydrocarbons can be subdivided into **alkanes**, **alkenes** and **alkynes**. Hydrocarbons for industrial use – for fuels, for lubricants and as starting points for a wide variety of industrial syntheses – are obtained principally from petroleum and coal tar. See **fractional distillation**.

> ### CONNECTIONS
>
> ORGANIC CHEMISTRY **82**
> HYDROCARBON CHAINS **84**
> CARBON-HYDROGEN COMPOUNDS **86**

hydrochloric acid

A solution of **hydrogen chloride** in water. The concentrated acid is about 35 percent hydrogen chloride and is corrosive. It is a typical strong, monobasic **acid** forming only one series of salts, the chlorides.

Hydrochloric acid has many industrial uses, including recovery of zinc from galvanized scrap iron and the production of chlorine, for example when oxidized by manganese(IV) oxide. It is also produced in the stomachs of animals for digestion.

hydrogenation

1 A reaction in which a hydrogen atom is incorporated into an **unsaturated** organic molecule (one that contains double bonds or triple bonds) to form a saturated one. Nickel is commonly used as a catalyst in such reactions. Hydrogenation is commonly used to turn vegetable oils into margarine. **2** The process of converting coal to oil by making the carbon in the coal combine with hydrogen to form hydrocarbons.

> ### CONNECTIONS
>
> ORGANIC CHEMISTRY **82**
> MAKING HYDROCARBONS **92**

hydrogen bond

A weak intermolecular (electrostatic) interaction between hydrogen atoms in associated compounds such as water, alcohols and proteins. See **bond**.

hydrogen carbonate

Also known as bicarbonate, a compound containing the hydrogen carbonate ion HCO_3^-, an acid salt of carbonic acid (a solution of carbon dioxide in water). When heated or treated with dilute acids, hydrogen carbonates give off carbon dioxide. The most important compounds are sodium hydrogen carbonate (bicarbonate of soda) and calcium hydrogen carbonate.

hydrogen chloride

A colorless, fuming gas (HCl). It is prepared in the laboratory by heating sodium chloride with concentrated sulfuric acid. On an industrial scale it is obtained as a byproduct of the chlorination of hydrocarbons, and directly from its constituent elements, both formed as byproducts in the manufacture of caustic soda. It is used in the manufacture of polyvinyl chloride (PVC) and other chlorine compounds. It dissociates fully in solution to form **hydrochloric acid**.

hydrogen ion

An atom of hydrogen carrying a positive charge (H^+). The concentration of hydrogen ions in solution is used as the basis for the measurement of **pH**.

hydrolysis

The chemical reaction in which the action of water or its ions breaks down a substance into smaller molecules. For example, salts of weak acids hydrolyze in aqueous solution. Hydrolysis occurs in certain inorganic salts in solution, in nearly all nonmetallic chlorides, in esters and in other organic substances. It is one of the mechanisms for the digestion of food by the body, as in the conversion of starch to glucose. Bodily hydrolysis reactions often need enzymes as catalysts.

hydrophilic

The term describing functional groups with a strong affinity for water, such as the carboxyl group (–COOH). If a molecule contains both a hydrophilic group and a group that repels water (see **hydrophobic**), it may have an affinity for both aqueous and nonaqueous molecules. Such compounds are commonly used as emulsion stabilizers (emulsifiers) or as **detergents**.

hydrophobic

The term describing functional groups that repel water. See also **hydrophilic**.

hydroxide

Any inorganic chemical compound containing one or more OH⁻ groups and generally combined with a metal. Hydroxides include sodium hydroxide (caustic soda, NaOH), potassium hydroxide (caustic potash, KOH) and calcium hydroxide (slaked lime, $Ca(OH)_2$). Metal hydroxides are usually basic in character. See also **alkali**.

hydroxyl group

An atom of hydrogen and an atom of oxygen bonded together (–OH) and covalently bonded to an organic molecule. Common compounds containing hydroxyl groups are alcohols and phenols. The hydroxyl group frequently behaves as a single entity in many chemical reactions.

hygroscopic substance

Any substance that absorbs water from the atmosphere but does not dissolve in it. A common example is copper(II) oxide.

hypochlorite

Any of a group of salts containing the ClO⁻ ion derived from hypochlorous acid (HOCl). Hypochlorites are made by absorbing chlorine into an alkali and are used for bleaching and as disinfectants.

ignition temperature

The minimum temperature to which a substance must be heated before it will spontaneously burn independently of the source of heat. For example, ethanol has an ignition temperature of 425°C.

indicator

A chemical compound that indicates when the concentration of a chemical species has passed a threshold value by change in color, turbidity or fluorescence. Indicators are used to find the end point in a **titration**. The commonest indicators detect changes in **pH** or in the oxidation state of a system. A universal indicator changes color continuously over a wide pH range.

INJECTION MOLDING

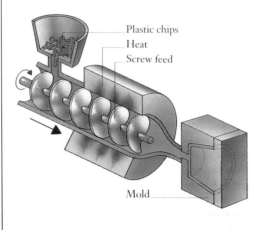

Plastic chips
Heat
Screw feed

Mold

inert gas
See **rare gas**.

injection molding
A method of forming a plastic into a specific shape. Molten plastic is injected from a heated cylinder by means of a screw or plunger into a water-cooled mold. It takes the shape of the mold as it solidifies.

inorganic chemistry
The branch of chemistry dealing with the elements and their compounds, excluding the more complex carbon compounds (*see* **organic chemistry**). The oldest known groups of inorganic compounds are acids, bases and salts. Another major group is the oxides, in which oxygen is combined with another element. Other groups are the compounds of metals with halogens (fluorine, chlorine, bromine, iodine and astatine), called halides (fluorides, chlorides and so on), and the compounds with sulfur (sulfides). All acids contain hydrogen. Acids containing one, two or three atoms of replaceable hydrogen are called mono-, di- or tri-basic, respectively. Salts are formed by the replacement of the acidic hydrogen of a polybasic acid by a metal or radical. If only part of the hydrogen is replaced, an acid salt is formed. Oxides are classified into: acidic oxides, forming acids with water; basic oxides, forming bases (containing the hydroxide group OH^-) with water; neutral oxides; and peroxides (containing more oxygen than the usual oxide). The basis of the description of the elements is the **Periodic Table** of elements.

CONNECTIONS
BONDS AND STRUCTURES **56**
ACIDS, BASES AND SALTS **62**

insulin
A hormone, produced by specialized cells in the islets of Langerhans in the pancreas, which regulates the metabolism of glucose, fats and proteins. Insulin was discovered by the Canadian physician Frederick Banting, who pioneered its use in treating **diabetes**. It was the first protein to have its structure determined completely.

intermolecular force
A force of attraction between molecules. Intermolecular forces are relatively weak; simple molecular compounds are gases, liquids or solids with low melting points. *See* **van der Waals' force** and **hydrogen bond**.

iodine
A grayish-black non-metallic element, with the symbol I, the atomic number 53, and relative atomic mass 126.9044. It is a **halogen**. When heated, its crystals give off a violet vapor with an irritating odor resembling that of chlorine. Iodine occurs only in combination with other elements. Its salts, known as iodides, are found in seawater. As a mineral nutrient it is vital to the proper functioning of the thyroid gland, where it occurs in trace amounts as part of the hormone thyroxine. Iodine is used in photography, in making dyes and in medicine as an antiseptic.

ion
An atom, or group of atoms, that is either positively charged (**cation**) or negatively charged (**anion**) as a result of the loss or gain of electrons during chemical reactions or through exposure to some forms of ionizing radiation.

ion exchange
The process by which the ions in one compound replace the ions in another. The exchange occurs because one of the products is insoluble in water. For example, when hard water is passed over an ion-exchange resin, the dissolved calcium and magnesium ions are replaced by sodium or hydrogen ions, so the hardness is removed. Commercial water softeners generally use ion-exchange resins. Adding washing-soda (sodium carbonate) crystals to hard water is also an example of ion exchange. The exchange of positively charged ions is called cation exchange; that of negatively charged ions is called anion exchange.

ion exchange chromatography
A type of **chromatography** in which the components of a mixture of ions in solution are separated according to the ease with which they will replace the ions on the polymer matrix through which they flow.

ionic bond
A chemical bond, also known as an electrovalent bond, produced when atoms of one element donate electrons to atoms of another element, forming positively- and negatively-charged ions, respectively. The electrostatic attraction between the oppositely charged ions constitutes the bond. Sodium chloride or common salt (Na^+Cl^-) is a typical ionic compound. *See* **bond**.

isomerism
The existence of chemical compounds (isomers) that have the same molecular composition and mass as each other, but with different physical or chemical properties due to the different arrangements of their atoms. For example, the organic compounds butane ($CH_3(CH_2)_2CH_3$) and methyl propane ($CH_3CH(CH_3)CH_3$) are isomers: each possesses four carbon atoms and ten hydrogen atoms but differ in the way that these are arranged. Structural isomers have obviously different constructions, but geometrical and optical isomers must be drawn or modeled in order to appreciate the difference in their three-dimensional arrangement. Geometrical isomers have a plane of symmetry and arise because of the restricted rotation of atoms around a bond; optical isomers are mirror images of each other. For instance, 1,1-dichloroethene ($CH_2=CCl_2$) and 1,2-dichloroethene ($CHCl=CHCl$) are structural isomers, but there are two possible geometric isomers of the latter, depending on whether the chlorine atoms are on the same side or on opposite sides of the plane of the carbon–carbon double bond.

CONNECTIONS
NATURAL POLYMERS **96**
MEDICAL DRUGS **114**
NATURAL DRUGS **116**

isotope
One of two or more atoms that have the same atomic number (same number of protons), but contain a different number of neutrons in their nuclei, thus differing in their atomic masses. They may be stable or radioactive, naturally occurring or synthesized. Hydrogen (one proton, no neutrons), deuterium (one proton, one neutron) and tritium (one proton, two neutrons) are isotopes of hydrogen. Most elements consist of a mixture of isotopes when in their natural state.

ketone
Any member of the group of organic compounds containing the carbonyl group ($=C=O$) bonded to two atoms of carbon

LIGAND

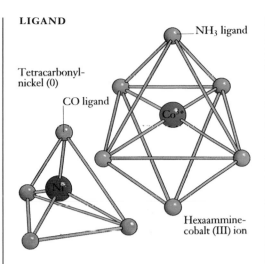

NH₃ ligand

Tetracarbonyl-
nickel (0)

CO ligand

Co³⁺

Ni

Hexaammine-
cobalt (III) ion

(*compare* **aldehydes**). Ketones are liquids or low-melting-point solids, slightly soluble in water. An example is propanone (acetone, CH_3COCH_3), used as a solvent.

lake dye
A pigment made by the interaction of an organic dyestuff with an inorganic compound such as an oxide, hydroxide or salt. Lake dyes are commonly found in paints and printing inks. *See* **anodizing** and **mordant**.

laminate
A composite material of two or more superimposed layers of different synthetic resin-impregnated or coated fillers. The layers may differ in their material, form or orientation. Bonding to form a single composite is usually by means of heat or pressure.

lanthanide
Any of a series of 15 metallic elements (also known as rare earths) set out in a band of the **Periodic Table** of elements ranging from atomic number 57 (lanthanum) to 71 (lutetium). One of its members, promethium, is radioactive. All occur in nature. Lanthanides are grouped because of their chemical similarities (they can all be divalent), their properties differing only slightly with increasing atomic number. Lanthanides were called **rare earths** originally because they were not widespread and were difficult to identify and separate from their ores. *See also* **actinide**.

latent image
The nondetectable image produced by the action of light on the silver halides in the emulsion of photographic film. The light causes the decomposition of the silver halide. The halide atom (usually bromine) is trapped in the gelatinous film base. The invisible particles of silver form the latent image, which is then made visible by the action of a **developer**.

ligand
A molecule or ion that donates a pair of electrons to a metal atom or ion when forming a coordinate bond. Monodentate ligands only have one point in each ligand where coordination can occur; polydentate ligands have two or more possible coordination sites. Many complexes with more than one kind of ligand have **stereoisomers**.

light
The form of electromagnetic radiation to which the human eye is sensitive. It has a wavelength from about 400 nm in the extreme violet to about 700 nm in the extreme red. Light is considered to exhibit particle and wave properties. Its fundamental particle, or quantum, is called a photon. The speed of light (and all electromagnetic radiation) in a vacuum is 299,792.5 km/sec (186,281 miles/sec) and is a universal physical constant.

liquid chromatography
A type of **chromatography** in which both the mobile phase (the mixture to the analyzed) and the stationary phases are liquid.

litmus
A dye obtained from various lichens that is used in chemistry as an indicator to test the acidic or alkaline nature of aqueous solutions. It turns red in the presence of acid and blue in the presence of alkali. *See* **test paper**.

lone pair
The name given to a pair of electrons in the outermost shell of an atom that are not used in bonding. In certain circumstances, they will allow the atom to bond with atoms, ions or molecules (such as boron trifluoride, BF_3) that are deficient in electrons, forming coordinate covalent (dative) bonds in which they provide both of the bonding electrons. *See* **bond**.

long-chain hydrocarbon
Any hydrocarbon with a linear molecule of high molecular weight.

macromolecule
A very large molecule, generally a **polymer**. Many natural and synthetic polymers contain macromolecules, as do substances such as the protein hemoglobin.

magnesium carbonate
A white solid compound ($MgCO_3$) that occurs in nature as the mineral magnesite. It is a commercial antacid, and the anhydrous form is used as a drying agent in table salt. When rainwater containing dissolved carbon dioxide flows over magnesite rocks, the carbonate dissolves to form magnesium hydrogen carbonate, one of the causes of temporary hardness in water.

magnesium oxide
A refractory, white, basic oxide, MgO, which occurs as the mineral periclase (magnesia). It is used as an antacid and, after conversion to the chloride, as a source of magnesium.

mass
The amount of substance that an object contains. It is the force of gravity acting upon mass that produces weight. As a result, the weight of an object on the Moon is less than on Earth, while its mass remains the same. Mass is effectively a measure of an object's inertia – in other words its resistance to acceleration. It is a scalar quantity; the international standard unit of mass is the kilogram.

mass number
The sum of the numbers of protons and neutrons in the nucleus of an atom. It is used along with the **atomic number** (the number of protons) in nuclear notation: in symbols that represent nuclear isotopes, such as $^{14}_{6}C$, the lower number is the atomic number, and the upper number is the mass number. The mass number is also known as the nucleon number.

mass spectrometer
A device for analyzing chemical composition. Positive ions of a substance are deflected through magnetic fields in a vacuum system on to a photographic plate, which permits accurate measurement of the relative concentrations of the various ionic masses present, particularly isotopes.

MASS NUMBER

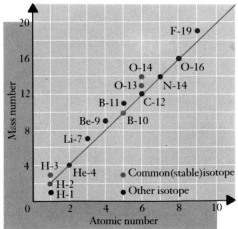

medical tests

Biochemical tests that are used to detect various substance in the body. For example, a number of illnesses can be detected as a result of abnormal levels of certain metabolites in the bloodstream. *See* **blood test**, **urine test** and **breath test**.

melamine

A white crystalline compound ($C_3N_6H_6$), made by heating dicyandiamide. It can be copolymerized with methanal to give thermosetting melamine resins, which are extremely resistant to heat and scratches. Such resins are used particularly for laminated coatings.

CONNECTIONS

TYPES OF PLASTIC **100**

SHAPING PLASTICS **102**

USING PLASTICS **104**

melting point

The temperature at which a substance melts, or changes from a solid to liquid form. A pure substance under standard conditions of pressure (usually one atmosphere) has a definite melting point. If heat is supplied to a solid at its melting point, the temperature does not change until the melting process is complete. The melting point of ice is 0°C.

mercury

A heavy, silver-gray, metallic element, with the symbol Hg (from Latin *hydrogyrum*), the atomic number 80, and the relative atomic mass 200.59. It is a dense, mobile liquid (also known as quicksilver) with a low melting point (–38.87°C). Its chief source is the mineral cinnabar (HgS), although it is sometimes found as a free metal. Its alloys with other metals are called **amalgams** (a silver-mercury amalgam is used in dentistry for filling cavities in teeth). Industrial uses include drugs and chemicals, mercury-vapor lamps, switches, barometers and thermometers. Mercury can accumulate in the human body and is poisonous.

metal

Any of a class of chemical elements with certain well-defined chemical characteristics and physical properties: they are good conductors of heat and electricity; opaque, but reflect light well; malleable, which enables them to be cold-worked and rolled into sheets; and ductile, which permits them to be drawn into thin wires. Metallic elements compose about 75 percent of the 106 elements shown in the **Periodic Table** of the elements. They can form alloys with each other; bases with the hydroxide ion (OH⁻) and replace the hydrogen in an acid to form a salt. Most metals are found in nature in the combined form only, as compounds or mineral ores; about 16 of them also occur in the elemental form, as native metals. Their chemical properties are largely determined by the extent to which their atoms can lose one or more electrons and form positive ions (cations).

CONNECTIONS

THE ELEMENTS **50**

MIXTURES AND COMPOUNDS **52**

ACIDS, BASES AND SALTS **62**

ELECTRICITY AND CHEMISTRY **72**

metalloid

Any chemical element having some of but not all the properties of metals. Metalloids are also known as semimetals and are usually electrical semiconductors. The elements silicon, germanium, arsenic, antimony and tellurium are metalloids.

methanal

Traditionally known as **formaldehyde**, a colorless gas (HCHO), condensing to a liquid at –21°C. It has a powerful, penetrating smell. Dissolved in water, it is known as formalin, used as a biological preservative. It is used in the manufacture of plastics, dyes, foam (for example, urea-formaldehyde foam, used in insulation) and in medicine.

methane

The simplest hydrocarbon (CH_4) of the alkane series. Colorless, odorless and lighter than air, it burns with a bluish flame and explodes when ignited with air or oxygen. Methane is the chief constituent of natural gas and also occurs in the explosive firedamp of coal mines. It causes about 38 percent of the warming of the globe through the greenhouse effect; the amount of methane in the air is predicted to double over the next 60 years. It has been estimated that 15 percent of all methane gas in the atmosphere is produced by cud-chewing animals.

methanoic acid

Also known as **formic acid**, a colorless, slightly fuming liquid (HCOOH) that freezes at 8°C and boils at 101°C. It is the simplest of the **carboxylic acids**. It occurs in stinging ants, nettles and pine needles, and is used in dyeing, tanning and electroplating.

methanol

Traditionally known as methyl alcohol, the simplest of the alcohols (CH_3OH). It can be made by the dry distillation of wood (it is also known as wood alcohol), but is usually made from coal or natural gas. When pure, it is a colorless, flammable liquid with a pleasant odor, and is highly poisonous. Methanol is used to produce methyl-ter-butyl ether (a replacement for lead as an octane-booster in gasoline), methanal, vinyl ethanoate (largely used in paint manufacture) and fuel for automobiles.

methyl group

The organic group $-CH_3$.

mineral

Any naturally formed inorganic substance with a defined chemical composition and a regularly repeating internal structure. Minerals are the constituents of rocks, either in their perfect crystalline form or otherwise. Minerals are generally classified by their anions in order of increasing complexity: sulfides, oxides, halides, carbonates, nitrates, sulfates, phosphates and silicates. In more general usage, a mineral is any substance economically valuable for mining (including coal and oil, despite their organic origins). Minerals are extracted from the Earth's crust by numerous different processes, including traditional open-cast mining, shaft mining and quarrying, as well as more specialized processes such as those used for oil and sulfur (*see* **Frasch process**). Mineral oil is oil obtained from mineral sources, such as coal or petroleum.

mixture

A system containing two or more compounds that still retain their separate physical and chemical properties. There is no chemical bonding between them. Unlike compounds, mixtures can be separated from each other by physical means, such as distillation, filtration or crystallization.

MINERAL

Zircon
(tetragonal)

Calcite
(trigonal)

Beryl
(hexagonal)

Kyanite
(triclinic)

Gypsum
(monoclinic)

Barytes
(orthorhombic)

mobile phase
See **chromatography**.

mole
SI unit (symbol mol) of the amount of a substance. It is defined as the amount of a substance that contains as many elementary units (atoms, molecules, ions, radicals, etc.) as there are atoms in 12 grams of the isotope carbon-12. One mole of an element that exists as single atoms weighs as many grams as its atomic number and contains 6.02253×10^{23} atoms (*see* **Avogadro's constant**).

molecular formula
The **chemical formula** indicating the actual number of atoms of each element present in a single molecule of a chemical compound. This is determined by two pieces of information: the empirical formula and the relative molecular mass, which is determined experimentally.

molecular weight
See **relative molecular mass**.

molecular mass
See **relative molecular mass**.

molecular modeling
The technique of determining the shapes, internal structures and properties of molecules based on knowledge of their constituent atoms and bond lengths. Molecular modeling can be used to develop synthetic drugs with very high specificity. Three-dimensional molecular modeling used to rely on ball-and-stick models, but the invention and application of low-cost, high-power computing has dramatically increased the scope and usefulness of molecular modeling techniques.

CONNECTIONS

BONDS AND STRUCTURES **56**

NAMES AND FORMULAS **60**

NATURAL DRUGS **116**

---

molecular sieve
Any porous crystalline substance, particularly an aluminosilicate, which can be dehydrated with very little change in its crystal structure. The resulting regularly-spaced cavities provide a high surface area for the adsorption of smaller molecules. Molecular sieves are used in the separation and purification of liquids and as drying agents. Because they can be loaded with chemicals that remain unreactive with them, they are also used as **catalysts** and catalyst support structures.

molecule
A group of two or more atoms bonded together. A molecule of an element consists of one or more like atoms, whereas a molecule of a compound consists of two or more different atoms bonded together. Molecules can vary in size and complexity from the hydrogen molecule (H_2) to the large macromolecules of proteins. They are held together by different types of **bonds**. The symbolic representation of a molecule is known as its formula. The presence of more than one atom is denoted by a subscript figure: for example, one molecule of the compound water, having two atoms of hydrogen and one atom of oxygen, is shown as H_2O. According to the molecular or kinetic theory of matter, molecules are in a state of constant motion, the extent of which depends on their temperature, and exert forces on one another. The shape of a molecule directly affects its chemical, physical and biological properties (*see* **isomerism**).

CONNECTIONS

MIXTURES AND COMPOUNDS **52**

TYPES OF BONDS **54**

BONDS AND STRUCTURES **56**

---

monomer
Any chemical compound composed of simple molecules from which **polymers** can be made. Under certain conditions the simple molecules (of the monomer) polymerize to form a very large macromolecule. Thus ethene is the monomer of the plastic polyethene (polyethylene*)*.

mordant
Any substance that combines with and fixes a dyestuff on a fiber that cannot be dyed directly. Mordants are usually the weak

MOLECULE

Oxygen (O_2)

Nitrogen (N_2)

Carbon dioxide (CO_2)

hydroxides of iron, aluminum and chromium. The color depends on the mordant used as well as the dyestuff. The action of a pigment on a mordant produces a **lake dye**.

naphthalene
A solid, white, volatile, aromatic hydrocarbon ($C_{10}H_8$) obtained from crude oil. The smell of mothballs is due to the napthalene content. It is used in making indigo and certain azo dyes, as a mild disinfectant and as an insecticide.

negative image
The image produced by the action of a **developer** on a **latent image** on photographic film. As a result of this process, the image is dark where a lot of light has fallen on the film and colorless where no light has fallen on the film. Light must be passed through the negative image onto photographic paper, where photochemical reactions similar to those involved in the previous stages of the photographic process produce a positive print. *See* **photography**.

neutralization
A process that occurs when the excess acid (or excess base) in a substance is reacted with added base (or added acid) so that the resulting substance is neither acidic nor basic; it is neutral, having a pH of 7.

neutron
One of the three chief **subatomic particles** (the others are the proton and the electron). Neutrons, which have about the same mass as protons but no electric charge, occur in the nuclei of all atoms except hydrogen. They contribute to the mass of atoms but do not affect their chemistry, which depends on the proton or electron numbers. **Isotopes** of a single element differ only in the number of neutrons in their nuclei and have identical chemical properties.

nitrate
Any salt or ester of nitric acid, containing the NO_3^- ion. Nitrates are widely used in explosives, in the chemical and pharmaceutical industries, and as fertilizers. Almost all nitrates are soluble in water, and only sodium and potassium nitrate occur significantly in nature. Nitrates play a major part in the **nitrogen cycle**.

nitration
The action of adding a nitro group ($-NO_2$) to an organic compound (or the substitution of another group for a nitro group). An example of nitration is the electrophilic substitution of benzene and related aromatic compounds using concentrated nitric acid.

nitric acid

Once known as *aqua fortis*, a fuming acid (HNO_3) obtained by the oxidation of ammonia or the action of sulfuric acid on potassium nitrate. It is a strong oxidizing agent and highly corrosive, dissolving most metals. HNO_3 is used in the nitration and esterification of organic substances, and in the making of sulfuric acid, nitrates, explosives, fertilizers, plastics and dyes.

nitrile

Any of a group of organic compounds, also known as cyanides, that contains the –CN group bound to an organic group. Nitriles are usually prepared by the reaction of potassium cyanide with haloalkanes in alcoholic solution, or by the dehydration of amides.

nitrite

Any salt or ester of nitrous acid that contains the nitrite ion (NO_2^-). Nitrites are used as preservatives (for example, to prevent the growth of the spores of the organism causing botulism) particularly in cured meats such as bacon and sausages.

nitrocellulose

A series of esters with two to six nitrate (NO_3^-) groups per molecule, made by the action of concentrated nitric acid on cellulose (such as cotton waste) in the presence of concentrated sulfuric acid. Those with five or more nitrate groups are explosive (gun cotton), but those with less were once used in lacquers, rayon (**artificial silk**) and plastics, such as colored and photographic film, until replaced by nonflammable cellulose acetate.

nitrogen

A gaseous, non-metallic element, with the symbol N, the atomic number 7, and the relative atomic mass 14.0067. Colorless, odorless and tasteless, it forms almost 80 percent of the Earth's atmosphere by volume and is a constituent of all plant and animal tissues (in proteins and nucleic acids). Nitrogen is obtained for industrial use by the liquefaction and fractional distillation of air. Its compounds are used to make foods, drugs, fertilizers, dyes and explosives. It plays an important role in plant nutrition through the **nitrogen cycle**. Nitrogen is used in the **Haber process** to make ammonia (NH_3) and to provide an inert atmosphere for certain chemical reactions.

CONNECTIONS

THE CHEMICAL INDUSTRY 74

VITAL RAW MATERIALS 110

NITROGEN CYCLE

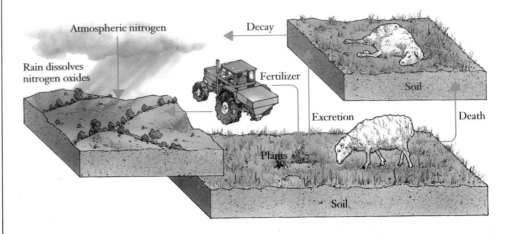

nitrogen cycle

The sequence of chemical reactions by which nitrogen circulates through the ecosystem. Nitrogen, in the form of inorganic compounds (such as nitrates) in the soil, is absorbed by plants and turned into organic compounds (such as proteins) in plant tissue. A proportion of this nitrogen is eaten by herbivores. Some in turn is passed on to carnivores that feed on the herbivores. The nitrogen is ultimately returned to the soil as excrement. When organisms die, it is converted back to inorganic form by decomposers before being taken in by plants again.

nitrogen fixation

The process by which nitrogen in the atmosphere is converted into nitrogenous compounds by the action of microorganisms in conjunction with certain legumes. The process indirectly makes nitrogen available to plants. Several chemical processes duplicate nitrogen fixation to produce fertilizers.

nitrogen oxide

Any chemical compound that contains only nitrogen and oxygen. All nitrogen oxides (NO_x) are gases. Nitrogen monoxide (nitric oxide) and nitrogen dioxide are exhaust products of internal-combustion engines and are thought to be involved in the process of depletion of the ozone layer, as well as contributing to general air pollution.

noble gas

See **rare gas**.

nonmetal

One of a set of about 20 elements with certain physical and chemical properties opposite to those of **metals**. They are typically poor conductors of heat and electricity, and seldom form positive ions. Because nonmetals accept electrons to form negative ions and covalent bonds, they are sometimes called electronegative elements. Nonmetals include carbon, nitrogen, oxygen, sulfur, phosphorus and the halogens.

CONNECTIONS

THE ELEMENTS 50

BONDS AND STRUCTURES 56

---

nuclear magnetic resonance (NMR)

The absorption of electromagnetic radiation of a precise frequency by a nucleus with a non-zero nuclear moment in an external magnetic field. The protons in water experience NMR at a radiation frequency of 12.6 MHz in a magnetic field of 0.3 tesla. Its main application is as NMR spectroscopy, which is widely used for the analysis of complex chemical structures.

nucleus

The positively-charged central part of an atom, that constitutes almost all its mass. Except for hydrogen nuclei, which have only protons, nuclei are composed of both protons and neutrons. Surrounding the nuclei are electrons, which have a negative charge equal to that of the protons, thus giving the atom a neutral charge.

nylon

Any of a group of synthetic long-chain polymers similar in structure to protein. Nylon was the first commercial all-synthetic fiber, made by United States chemist Wallace Carothers from petroleum, natural gas, air and water in 1938. There are three main nylon fibers: nylon 6, nylon 6,6 and nylon 6,10, formed by different polymerization schemes. All are strong, elastic and relatively insensitive to moisture, and are used in molding, textiles and medical sutures.

oil

A flammable substance, usually insoluble in water, and composed chiefly of carbon and hydrogen. Oils may be solids (fats and waxes) or liquids, and are categorized as fixed or volatile according to the ease with which they vaporize when heated. Essential oils are volatile liquids that are obtained from plants and are used in perfumes, flavoring essences and in aromatherapy. Fixed oils are mixtures of lipids, of varying consistency, found in both animals (for example, fish oils) and plants (in nuts and seeds); they are used as foods and as lubricants, and in the making of soaps, paints and varnishes. Mineral oils, obtained chiefly from the refining of petroleum, are composed of a mixture of hydrocarbons, and are used as fuels and lubricants.

CONNECTIONS

olefin

See **alkene**.

optical activity

The ability of certain crystals, liquids and solutions to rotate the plane of plane-polarized light as it passes through them. The phenomenon is related to the three-dimensional arrangement of the atoms making up the molecules concerned. It occurs when the molecules of the substance are asymmetric, so that they can exist in two structural forms, each being a mirror image of the other. The two forms are known as optical isomers or **enantiomers**.

optical brightener

Any compound that, when added to a white textile material, increases its apparent brightness by converting some of the ultraviolet radiation into visible blue light. Optical brighteners are also known as fluorescent whitening agents.

orbital

The region around the nucleus of an atom (or, in a molecule, around several nuclei) in which an electron is most likely to be found. According to quantum theory, the position of an electron is uncertain; it may be found at any point. However, it is more likely to be found in some places than in others, and it is these places that make up the orbital. An atom or molecule has numerous orbitals, each of which has a fixed size and shape. An orbital is characterized by three numbers, called quantum numbers, representing its energy (and hence size), its angular momentum (and hence shape), and its orientation. Each orbital can be occupied by one or (if their spins are oppositely aligned) two electrons. In a **covalent bond**, molecular orbitals are formed by the linear combination of the outer atomic orbitals of the atoms.

ore

Any naturally occurring body of rock or deposit of sediment from which metal may be extracted commercially. Sometimes metals are found uncombined (native metals), but more often they occur as compounds such as carbonates, sulfides or oxides. The ores often contain unwanted impurities that must be removed when the metal is extracted. Commercially valuable ores include bauxite (aluminum oxide, Al_2O_3) hematite (iron(III) oxide, Fe_2O_3), zinc blende (zinc sulfide, ZnS) and rutile (titanium dioxide, TiO_2).

organic chemistry

The branch of chemistry that deals with carbon compounds. In a typical organic compound, each carbon atom forms bonds covalently with each of its neighboring carbon atoms in a chain or ring, and additionally with other atoms, commonly hydrogen, oxygen, nitrogen or sulfur. The basis of organic chemistry is the ability of carbon to form long chains of atoms, branching chains, rings and other complex structures. Compounds containing only carbon and hydrogen are known as **hydrocarbons**. Organic chemistry is largely the chemistry of a great variety of **homologous series**. The linking carbon atoms that form the backbone of an organic molecule may be built up from beginning to end without branching; or may throw off branches at one or more points. Sometimes, however, the chains of carbon atoms can also form rings (**cyclic compounds**), usually of five, six or seven atoms. Open-chain and cyclic compounds may be classified as **aliphatic** or **aromatic** depending on the nature of the bonds between their atoms. Compounds containing oxygen, sulfur, or nitrogen within a carbon ring are called heterocyclic compounds. Many organic compounds exhibit **isomerism**.

CONNECTIONS

oxidation

The loss of electrons, gain of oxygen or loss of hydrogen by an atom, ion or molecule during a chemical reaction. Oxidation may be brought about by reaction with another compound (oxidizing agent), which simultaneously undergoes reduction, or it may occur electrically at the anode (positive electrode) of an electrolytic cell. *See also* **redox reaction.**

oxide

Any compound of oxygen and another element, often produced by burning the element or a compound of it in air or oxygen. Oxides of metals are normally bases and react with an acid to produce a salt in which the metal forms the cation. Some of them also react with a strong alkali to produce a salt in which the metal is part of a complex anion. Most of the oxides of nonmetals are acidic (dissolve in water to form an acid). Some oxides display no pronounced acidic or basic properties – they are neutral.

oxygen

A nonmetallic, gaseous element, with the symbol O, the atomic number 8, and the relative atomic mass 15.9994. Colorless, odorless and tasteless, it is the most abundant element in the Earth's crust (almost 50 percent by mass), forms about 21 percent by volume of the atmosphere, and is present in combined form in water and many other substances. Life on Earth evolved using oxygen, which is a byproduct of photosynthesis and the basis for respiration in plants and animals. Oxygen is very reactive and combines with all other elements except the inert gases and fluorine. In nature it normally exists as a molecule composed of two atoms (O_2); single atoms of oxygen are very short-lived owing to their reactivity (*see* **free radical**). They can be produced in electric sparks and by the Sun's ultraviolet radiation in space, where they rapidly combine with molecular oxygen to form **ozone** (O_3). Oxygen is obtained for industrial use by the fractional distillation of liquid air, by the electrolysis of water or by heating manganese(IV) oxide with potassium chlorate.

ozone

A highly reactive pale-blue gas (O_3) with a penetrating odor. Ozone is an allotrope of oxygen (*see* **allotropy**), made up of three atoms of oxygen. It is formed naturally when the molecule of the stable form of oxygen (O_2) is split by ultraviolet radiation or electrical discharge. Ozone forms a thin layer in the upper atmosphere, which protects life on Earth from damaging ultraviolet rays. At lower atmospheric levels it is an air pollutant

OZONE HOLE

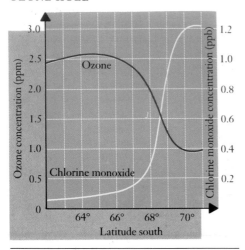

and contributes to the greenhouse effect. Near the ground, higher-than-usual concentrations of ozone can cause asthma attacks, stunted growth in plants and corrosion of certain materials. It is produced by the action of sunlight on air pollutants (*see* **smog**). Ozone is a powerful oxidizing agent and is used industrially in bleaching and for sterilization in water or air conditioning.

ozone hole
A decrease in the concentration in the thin layer of ozone in the stratosphere by any of a number of ozone depleters – chemically stable compounds containing chlorine or bromine, which remain unchanged for long enough to drift up to the upper atmosphere. The best known are **chlorofluorocarbons** (CFCs), but many other ozone depleters are known, including **halons**, used in some fire extinguishers; trichloroethane and tetra-chloro-methane, both solvents; some substitutes for CFCs; and the pesticide bromomethane. The extent of ozone depletion is out of all proportion to the amount of the depleter in the upper atmosphere, because a chain reaction can be set up by a single CFC molecule that results in the destruction of many thousands of ozone molecules.

"Holes" in the ozone layer are dangerous because they allow large quantities of ultraviolet light, a major cause of skin cancer, to reach the Earth's surface. Ozone depletion over the polar regions is the most dramatic manifestation of a general global effect.

paint
A suspension of pigments in a liquid vehicle (binder and solvents) which, when applied to a surface, dries to give a hard, adhesive coat for decoration and protection. Common types of paint are cellulose paints (lacquers), oil-based paints, emulsion paints and specialized paints such as primers and enamels.

paper–liquid chromatography
See **chromatography**.

paraffin
See **alkane**.

periodicity
The basis of the **Periodic Table**, by which elements are arranged according to their atomic number, with certain chemical and physical properties reappearing at fixed intervals.

Periodic Table
A table in which the elements are arranged in order of their atomic number (*see* **periodicity**). The table summarizes the major properties of the elements and enables predictions to be made about their behavior. There are striking similarities in the chemical properties of the elements in each of the vertical columns (called **groups**), which are numbered I-VIII and 0. The groups reflect the number of electrons in the outermost unfilled electron **shell** and hence the maximum **valence**. A gradation of properties may be traced along the horizontal rows (called periods). Metallic character increases across a period from right to left, and down a group. A large block of elements, between Groups II and III, contains the **transition elements**, characterized by displaying more than one valence state. These features are a direct consequence of the electronic (and nuclear) structure of the atoms of the elements. The relationships established between the positions of elements in the Periodic Table and their major properties has enabled scientists to predict the properties of other elements; for example, technetium, atomic number 43, which was first produced in the laboratory in 1937. The first Periodic Table was devised by the Russian chemist Dmitri Mendeleyev in 1869.

CONNECTIONS

THE ELEMENTS 50

MIXTURES AND COMPOUNDS 52

TYPES OF BONDS 54

---

petrochemicals
Chemicals derived from the processing of petroleum (crude oil) or natural gas. Petrochemical industries are those that obtain their raw materials from the processing of petroleum and natural gas. Polymers, detergents, solvents and nitrogen fertilizers are all major products of the petrochemical industries. Inorganic chemical products include carbon black, sulfur, ammonia and hydrogen peroxide.

petroleum
A natural mineral oil, also known as crude oil, found underground in permeable rock as a thick, greenish-brown flammable liquid. Petroleum consists of hydrocarbons mixed with oxygen, sulfur, nitrogen and other elements in varying proportions. It is thought to be derived from ancient organic material that has been converted by, first, bacterial action, then heat and pressure (but its origin may be chemical also). Various products are made from crude petroleum by **fractional distillation** and other processes. Petroleum products and chemicals are used in large quantities in the manufacture of pharmaceuticals, synthetic fibers, plastics, fertilizers, detergents, toiletries, and synthetic rubber. The burning of petroleum fuel is one cause of air pollution. *See* **exhaust emission**.

pH
A logarithmic scale from 0 to 14 for measuring acidity or alkalinity. pH stands for "potential hydrogen". At 25°C, a pH of 7.0 indicates neutrality; below 7 is acid; while a pH above 7 is alkaline. The pH value of an aqueous solution equals the negative logarithm of the concentration of hydrogen ions. pH can be measured by using a broad-range indicator or a pH meter.

pharmacology
The study of drugs, their chemistry, their mode of action, interactions and side effects.

phase
A physical state of matter: for example, steam, ice and liquid water are the three different phases of water; a mixture of two is termed a two-phase system.

phenol
Any member of a group of aromatic chemical compounds with weakly acidic properties, which are characterized by a hydroxyl (–OH) group attached directly to an aromatic ring. Unlike alcohols, phenols are acidic because of the influence of the aromatic ring. The simplest of the phenols, derived from benzene, is also known as phenol and has the formula C_6H_5OH.

phenylethene
A colorless, hydrocarbon ($C_6H_5CH=CH_2$), also known as **styrene**, soluble in ethanol and ethoxyethane. It is used in the production of **polystyrene**.

phosphate
A salt or ester of phosphoric acid (H_3PO_4). Incomplete neutralization of phosphoric acid creates acid phosphates. Phosphates are used as fertilizers and are involved in many

PHOTOGRAPHY

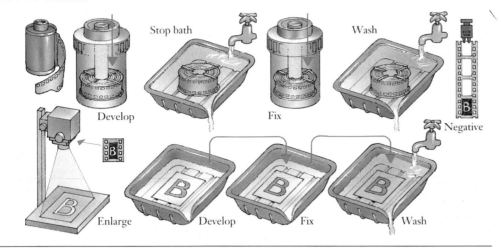

Stop bath

Wash

Develop

Fix

Negative

Enlarge

Develop

Fix

Wash

biochemical processes, often as part of complex molecules, such as adenosine triphosphate. Accumulated phosphates from fertilizers and other sources (such as detergents) have polluted many aquatic systems.

phosphoric acid
A white crystalline solid derived from phosphorus and oxygen. Its commonest form (H_3PO_4), also known as orthophosphoric acid, is produced by the action of phosphorus pentoxide (P_2O_5) on water. It absorbs water easily and is used as a concentrated aqueous solution. It is the most commercially important derivative of phosphorus, and is used to remove and prevent rust.

photochemistry
The area of chemistry concerned with the effects of radiation, particularly in the visible and ultraviolet regions of the electromagnetic spectrum. Photochemical reactions include those that produce light, and those that are initiated by light. One of the first photochemical reactions to be explained fully was the production of hydrogen chloride from a mixture of gaseous hydrogen and chlorine exposed to bright light. In this case the molecules of gas absorb light energy and split to form individual atoms of H and Cl (initiation). The individual atoms can then recombine with dissimilar atoms to produce HCl (propagation) or similar atoms to reform the initial gases (termination). Photochemistry is important in photography, photosynthesis and in the study of dyes.

photodissociation
Dissociation of a compound caused by the absorption of light energy, which usually results in the formation of **free radicals**. It is responsible for many of the harmful substances found in photochemical smog.

photography
The photochemical reproduction of images on sensitized materials. The most common photographic techniques use an acetate film base coated with light sensitive silver halides in gelatin. The silver halides are activated by light to form a **latent image** of silver. This latent image is then processed to a **negative image** by use of a **developer**. Color photography makes use of various dyestuffs that are added to different layers of silver halides to make them sensitive to the different primary colors.

photon
The elementary particle (package) of energy in which light and other forms of electromagnetic radiation are emitted. The photon has both particle and wave properties. It has no charge and is considered massless, but it possesses momentum and energy. It carries electromagnetic force (one of the fundamental forces of nature). A photon is emitted whenever an electron in an atom loses energy; when absorbed by an atom it excites an electron into a higher energy state.

photosynthesis
The biochemical process by which green plants utilize light energy to drive a series of chemical reactions, leading to the formation of carbohydrates. Photosynthesis requires the presence of chlorophyll, and the plant must have a supply of carbon dioxide and water. Actively photosynthesizing green plants store excess sugar as starch. The chemical reactions of photosynthesis occur in two stages. During the light reaction sunlight is used to split water (H_2O) into oxygen (O_2), hydrogen ions (H^+) and electrons, and oxygen is given off as a byproduct. In the dark reaction, for which sunlight is not required, the protons and electrons are used to convert carbon dioxide (CO_2) into carbohydrates ($C_n(H_2O)_m$). Photosynthesis depends on the ability of chlorophyll to capture the energy of sunlight and to use it to split water molecules. Other pigments, such as carotenoids, are also involved in capturing light energy and passing it on to chlorophyll.

pigment
Any synthetic or naturally occurring insoluble colored material that, when mixed in a suitable medium, is able to impart color. *See* **dye** and **paint**.

plasticizer
Any substance added to a synthetic polymerso as to preserve or enhance its flexibility. Commonly used plasticizers include triphenyl phosphate, tricresyl phosphate and high-boiling glycol esters.

plastic
A general term for any material that can be molded into different shapes. Rubber, bitumen and resins are naturally occurring plastics, but most plastics are synthetic polymers derived from petroleum. A molecule of plastic has thousands or millions of atoms, based on a carbon backbone. Plastics can be tailored in the laboratory to suit specific purposes – for instance, they can be made hard, elastic, transparent, tough, inert and resistant to water and corrosion. The two main kinds of plastics are **thermoplastics**, which soften when warmed and harden when they cool, and **thermosetting plastics**, which become rigid when set and do not soften again on warming. Shape-memory polymers are plastics that can be crumpled or flattened and will resume their original shape when heated.

polar bond
A covalent bond between two elements of different electronegativity. *See* **bond**.

polyaddition
Multiple **addition reactions** in a polymerization process. *See* **polymerization**.

polyamide
A type of natural or synthetic condensation polymer with a structure similar to that of proteins. Polyamides are formed by the interaction of an amine group from one molecule with a carboxylic acid group of another molecule (forming an amide group –CONH–). The chains of polyamide are linked together by hydrogen bonds. Examples of natural polyamides are silk, hair and wool; synthetic polyamides include nylon. Polyamides with aromatic groups attached to the amide groups are known as aramid fibers.

polycarbonate
Any of a range of thermoplastics produced by the condensation of phosgene (carbonyl chloride) with dihydroxy organic compounds, such as diphenylol propane. Polycarbonates are tough, clear, slightly tinted materials that, although expensive, are used in shatterproof and bulletproof windows.

polycondensation
Multiple **condensation reactions** in a polymerization process. *See* **polymerization**.

polyethene
A tough, white, translucent, waxy thermoplastic also known as polyethylene. It is polymer of the gas ethene C_2H_4. There are two types of normal polyethene: low-density polyethene (LDPE) made by high-pressure polymerization of ethene gas, and high-density polyethene (HDPE), which is made at lower pressure by using catalysts, and is more rigid at low temperatures and softer at higher temperatures than the low-density type. It has a variety of uses in manufacturing, packaging and as an electrical insulator.

polyethylene
See **polyethene**.

polymer
Any compound of large molecules made up of a long-chain or branching matrix composed of repeated simple units (**monomers**). Many types of polymer exist, both natural (proteins, nucleic acids, polysaccharides and many minerals) and synthetic (polyethene, nylon and plastics). There are two main groups of synthetic polymer: **thermoplastic** and **thermosetting**.

polymer composite
A composite material made up of two or more different types of polymer.

polymerization
A chemical reaction in which of two or more (usually small) molecules of the same kind join together to form a new compound. Addition polymerization (polyaddition) produces simple multiples of the same compound. In this type of polymerization, many molecules of a single compound join together to form long chains. An example is the polymerization of ethene to polyethene. Other addition polymers include polyvinyl chloride (PVC). In condensation polymerization (polycondensation), a small molecule such as water or hydrogen chloride is given off as a result of the polymerization. One example is the production of polyesters, formed by the polymerization of organic acids and alcohols. Condensation polymerization may involve a single monomer that has two reactive groups, such as an amino acid, or two or more different monomers, such as urea and methanal (formaldehyde); in this case, the polymerization is known as copolymerization.

polystyrene
A clear, glasslike plastic produced by the free radical polymerization of phenylethene. Polystyrene is used as a thermal and electrical insulator, for packing and decoration.

polyvinyl chloride (PVC)
A tough white solid plastic that softens with the application of a plasticizer. It is produced from chloroethene and is also known as polychloroethene; it is the best known and most widely used vinyl plastic. It is easy to color and resistant to fire, weather and chemicals, and has a variety of uses including drainpipes, floor tiles, audio disks and shoes.

potassium
A soft, waxlike, silver-white, metallic element, symbol K (Latin *kalium*), atomic number 19, relative atomic mass 39.0983. It is one of the alkali metals, has a very low density and is the second lightest metal (after lithium). It oxidizes rapidly when exposed to air and reacts violently with water. It is widely distributed in the Earth's crust with other elements, and is found in salt and mineral deposits in the form of potassium aluminum silicates. The element was discovered by the English chemist Humphry Davy, who isolated it from potash in 1807, in the first instance of a metal being isolated by **electrolysis**.

precipitate
An insoluble solid formed in a liquid as a result of a reaction within the liquid between two or more soluble substances. Precipitates usually settle out of solution; however, if the particles are very small, they may remain in suspension, forming a colloidal precipitate (*see* **colloid**).

pressure
The physical force acting at right angles to a body per unit surface area. The SI unit of pressure is the pascal (newton per square meter) equal to 0.01 millibars. In a fluid (liquid or gas), pressure increases with depth. At sea level atmospheric pressure is about 100 kilopascals (1013 millibars or 1 atmosphere).

POLYMERIZATION

Addition polymerization

Ethene monomers

Polyethene

Condensation polymerization

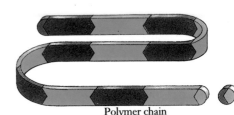

Monomer   Monomer

Dimer   HCl

Polymer chain

product yield

The amount of **substrate** actually produced in a chemical reaction expressed as a percentage of the theoretical amount of substrate that should be produced, calculated based on knowledge of the chemical natures of the original reactants.

propanone

Traditionally known as acetone, a colorless flammable volatile liquid (CH_3COCH_3) used extensively as a solvent and as a raw material for making plastics. Propanone is the simplest ketone. It boils at 56.5°C, mixes with water in all proportions and has a characteristic odor.

protein

The general name given to any complex, biologically important polymeric substance composed of amino acids joined by peptide bonds. Other types of bond are responsible for creating the protein's characteristic three-dimensional structure, which may be fibrous, globular or pleated. Proteins are essential to all living organisms. As enzymes they regulate all aspects of metabolism. Structural proteins such as keratin and collagen make up skin, claws, bones, tendons and ligaments; muscle proteins produce movement; hemoglobin transports oxygen; and membrane proteins regulate the movement of substances into and out of cells.

> ### CONNECTIONS
>
> CHEMISTRY OF LIFE **108**
> VITAL RAW MATERIALS **110**
> LIVING CHEMISTRY **112**

proton

A positively-charged **subatomic particle**, a fundamental constituent of any atomic nucleus. A proton carries a unit positive charge equal to the negative charge of an electron. Its mass is almost 1,836 times that of an electron, or 1.67×10^{-24} grams. The number of protons in the atom of an element is equal to its **atomic number**.

pyrolysis

The decomposition of a substance by heating it to a high temperature in the absence of air. The process is used to burn and dispose of old tires, for example, without contaminating the atmosphere. The term is also a synonym for **cracking**.

qualitative analysis

A procedure for determining the chemical identity of the components of a single substance or mixture. Qualitative analysis usu-ally involves a series of simple reactions and tests on the compound to determine the elements present. Chromatography or spectroscopy may also be used.

quantitative analysis

A chemical procedure for determining the precise amount of a known component present in a single substance or mixture. A known amount of the substance is subjected to particular procedures. Gravimetric analysis determines the mass of each constituent present; volumetric analysis determines the concentration of a solution by **titration** against a solution of known concentration.

> ### CONNECTIONS
>
> CHEMICAL ANALYSIS **134**
> VOLUMETRIC ANALYSIS **138**

quartz

See **silica**.

racemic mixture

An optically inactive mixture of equal quantities of the mirror image forms of optical isomers. Such mixtures are produced during the synthesis of compounds with asymmetric molecules from optically inactive starting materials. The optically active compounds can be isolated by a number of methods, such as coupling with an optically active substance or by the action of various bacteria and yeasts which only attack one isomer. *See* **optical activity**.

radioactivity

The spontaneous decay of the nuclei of radioactive atoms, accompanied by the emission of radiation. Radioactivity is exhibited by the radioactive isotopes of elements that are normally found in stable form and all iso-topes of radioactive elements. It may be either natural or induced. Radioactivity establishes an equilibrium in parts of the nuclei of unstable radioactive substances, ultimately forming a stable arrangement of nucleons (protons and neutrons). This is most frequently accomplished by the emission of alpha particles (helium nuclei, comprising two protons and two neutrons), beta particles (electrons and positrons) or gamma radiation (electromagnetic waves of very high frequency). It takes place either directly, through a one-step decay, or indirectly, through a number of decays that transmute one element into another. The instability of the particle arrangements in the nucleus of a radioactive atom determines the lengths of the half-lives of the isotopes of that atom. *See* **half-life**.

radioisotope

A synthetic or naturally occurring radioactive form of an element. Most radioisotopes are produced by bombarding a stable element with neutrons in the core of a nuclear reactor. The radiation given off is easy to detect (radioisotopes are also used as tracers). Radioisotopes are used in the fields of medicine, industry, agriculture and research.

rare-earth element

See **lanthanide**.

rare gas

Also known as noble gas, any of a group of six elements (helium, neon, argon, krypton, xenon and radon) in Group 0 of the Periodic Table of elements. Rare gases were also known as inert gases because they were originally thought not to enter into any chemical reactions, although this is now known to be incorrect. The extreme unreactivity of these gases is due to the stability of their electronic structure. All the electron shells of inert gas atoms are full. All except helium have eight electrons in their outermost (valence) shell.

rayon

A synthetic fiber produced from cellulose (originally in the form of wood pulp). There are two types of rayon, viscose rayon and acetate rayon, which differ in the method of production of the cellulose filaments.

reactant

Material that takes part in a chemical reaction. Reactants are written on the left-hand side of the **equation** describing a reaction.

reaction, chemical

The interaction of two or more atoms, ions or molecules (reactants), resulting in chemically different compounds (products). All

RADIOACTIVITY

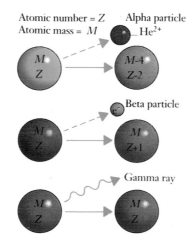

Atomic number = Z
Atomic mass = M

Alpha particle
He^{2+}

M Z → M-4 Z-2

Beta particle

M Z → M Z+1

Gamma ray

M Z → M Z

reactions are reversible to some extent; the products can react to give rise to the original reactants (*see* **equilibrium**). However, in many cases this back reaction is virtually nonexistent and the reaction is irreversible.

reaction rate
The rate at which a chemical reaction proceeds; that is, the reactants are used up or the products are formed. In a reaction that includes a number of steps, the slowest step determines the overall rate of reaction.

reaction kinetics
The study of the various intermediate stages involved in a chemical reaction.

reactivity
1 An imprecise term that describes the ease with which an element or compound takes part in a chemical reaction. For example, the **alkali metals** (sodium, potassium, etc.) and the **halogens** (fluorine, chlorine, etc.) have a high reactivity, whereas the **rare gases** (helium, argon, etc.) have extremely low reactivity. **2** The ease with which a reactive dye forms covalent chemical bonds with a fiber.

> ### CONNECTIONS
> THE ELEMENTS **50**
> TYPES OF BOND **54**
> BONDS AND STRUCTURES **56**
> ACIDS, BASES AND SALTS **62**

recycling
The processing of household or industrial waste (such as some metals and plastics) so that it can be reused, thus saving expenditure on scarce raw materials and helping to reduce environmental pollution. In many commercial chemical processes, the byproducts of the main reactions are recycled

RECYCLING

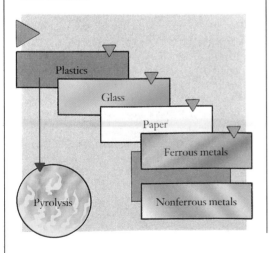

directly in the reaction. For example, all intermediate byproducts are recycled in the ammonia–soda process for the production of sodium carbonate.

> ### CONNECTIONS
> A CLEANER ENVIRONMENT **70**
> MAKING USEFUL CHEMICALS **78**
> RECYCLING POLYMERS **106**

redox reaction
A reduction–oxidation reaction in which one of the reactants is reduced and the other is oxidized at the same time. The reaction occurs only if both of the reactants are present and each changes simultaneously. For example, hydrogen reduces copper(II) oxide to copper while it is oxidized to water. The corrosion of iron and the reactions taking place in electrolysis are also examples of redox reactions.

reducing agent
Any substance that can reduce another substance (*see* **reduction**). In a redox reaction, it is the reducing agent that is itself oxidized. Strong reducing agents include hydrogen, carbon monoxide, carbon and metals. Reducing agents are also called reductants.

reduction
The process in which an atom, ion or molecule gains electrons, loses oxygen or gains hydrogen during a chemical reaction. Reduction may be brought about by reaction with another compound, which is simultaneously oxidized (reducing agent), or electrically at the cathode (negative electrode) of an electric cell. Examples include the reduction of iron(III) oxide to iron by carbon monoxide, the hydrogenation of ethene to ethane and the reduction of a sodium ion to sodium.

refining
The production of pure materials from impure materials or mixtures of materials. For example, petroleum is refined into useful hydrocarbons by **fractional distillation** and **cracking**. Electrolysis and flotation are other commonly used methods of refining.

relative atomic mass (RAM)
The mass of an atom of an element relative to one-twelfth the mass of an atom of carbon-12. It depends only on the number of protons and neutrons in the atom. If more than one isotope of the element is present, the relative atomic mass is based on the relative proportions of each isotope, resulting in values that are not whole numbers. The relative atomic mass is given on the unified

scale, where 1 atomic mass unit = 1.660×10^{-27} kg. The term **atomic weight**, although commonly used, is technically incorrect.

relative molecular mass
The mass of a molecule calculated relative to one-twelfth the mass of an atom of carbon-12. It is found by adding the relative atomic masses of the constituent atoms that make up the molecule. The term **molecular weight** is technically incorrect.

resin
Any substance produced by polymerization (more precisely, it is applied to a polymeric compound before setting). Natural resins are acidic and secreted by many trees; they are found as brittle glassy substances or dissolved in natural oils. Synthetic resins are used in adhesives, plastics and paints.

rubber
Natural rubber is a polymeric substance obtained from the sap of a variety of plants, particularly the tree *Hevea brasiliensis*. The coagulated and dried sap (latex) is modified by vulcanization (heating with sulfur or sulfur compounds) and compounding with fillers. *See also* **synthetic rubber**.

> ### CONNECTIONS
> NATURAL POLYMERS **96**
> SYNTHETIC POLYMERS **98**

salt
Any member of a group of compounds containing a positive ion (cation) derived from a metal or ammonia and a negative ion (anion) derived from an acid or nonmetal. If the negative ion has a replaceable hydrogen atom, it is an acid salt (for example, sodium hydrogen carbonate, $NaHCO_3$); if not, it is classed as a normal salt (for example, sodium chloride, NaCl). Salts have the properties typical of ionic compounds.

> ### CONNECTIONS
> MIXTURES AND COMPOUNDS **52**
> TYPES OF BONDS **54**
> ACIDS, BASES AND SALTS **62**

saponification
The splitting (hydrolysis) of an ester by treatment with a strong alkali, producing the alcohol from which the ester had been derived and a salt of the constituent fatty acid. The process is used in making soap, in which the soap is salted out by adding a saturated solution of sodium chloride.

saturated

The term given to an organic compound, such as propane, that contains only single covalent bonds. Saturated organic compounds can only undergo further reaction by substitution reactions, as in the production of chloropropane from propane, but not by addition reactions. Solutions in which no more solute can dissolve are also said to be saturated.

saturated fat

See **fat**.

scrubber

A piece of equipment used in industrial processes to remove certain gases (by dissolving them). For example, one type of scrubber removes sulfur dioxide (SO_2) from furnace flue gases, so preventing it from becoming an atmospheric pollutant and potential cause of acid rain.

shell

A grouping of electrons around an atom. The inner shell can hold two electrons, the next shell eight, and so on, up to 32 for the outer shells. A full shell is more stable than an incomplete one: thus atoms with "full"

SALT

Neutralization and evaporation

Precipitation and filtration

Dissolution in acid and evaporation

outer shells tend to be unreactive, whereas those with almost full or almost empty outer shells tend to be highly reactive, forming bonds with other atoms which permit these shells to be filled, by sharing electrons.

silica (silicon dioxide)

A white or colorless crystalline mineral (SiO_2) that is insoluble in water, but soluble in hydrofluoric acid and strong alkali. In its natural state it can exist in four forms: cristobalite, tridymite, quartz and lechatelierite. It is one of the most abundant minerals of the Earth's crust (12 percent by volume). Silica is widely used in ordinary glass, glazes and enamels, and as silica brick, a highly refractive furnace lining.

silicate

Any compound containing silicon and oxygen, combined together as the negative ion (anion), and one or more metal cations. A large number of high molecular mass, complex silicates exist, all containing the basic structural unit of the SiO_4 tetrahedron. Common natural silicates are sands (yellow sand is a type of **silica** containing iron oxide impurities). Glass is a manufactured complex polysilicate material in which other elements (such as boron) have been incorporated.

silicon

A metalloid element, with the symbol Si, the atomic number 14 and the relative atomic mass 28.08. Silicon is the second most abundant element, after oxygen, occurring as silica and silicates. Silicon forms an amorphous brown powder, or gray semiconducting crystals. It is used in alloys and to make transistors and semiconductors.

silver halide

Any of a set of compounds formed between silver and halogens. Common silver halides include silver bromide, silver chloride and silver iodide. All of these halides can be precipitated from silver nitrate solution by the addition of the appropriate halogen ions. Silver bromide and silver chloride are commonly used in photographic emulsions; on exposure to light and after development they are reduced to silver. *See* **photography**.

single bond

See **covalent bond** and **carbon–carbon bond**.

smog

A polluting mixture of smoke, fog and chemical fumes. Photochemical smog is produced under certain climatic conditions by the complex photochemical reaction of sunlight with unburned hydrocarbons, particularly

from exhaust emissions, which enter the upper atmosphere above the Earth.

CONNECTIONS

A CLEANER ENVIRONMENT 70
ATMOSPHERIC CHEMISTRY 132

soap

Any salt of a metal and various fatty acids (the most common being palmitic, stearic and oleic acid). Water-soluble soaps are made by the action of sodium hydroxide or potassium hydroxide. Soap disperses grease and dirt in water; however, unlike detergents, soaps form insoluble salts with magnesium and calcium ions present in hard water to produce a scum.

soda

See **sodium carbonate** and **sodium hydroxide**.

sodium carbonate

Also known as soda ash, an anhydrous white solid (Na_2CO_3). The hydrated, crystalline form ($Na_2CO_3.10H_2O$) is also known as washing soda. It is usually made by the Solvay process (*see* **ammonia–soda process**) and is used as a mild alkali, because it is hydrolyzed in water. Sodium carbonate is widely used in glass manufacture, in water softening, in textile treatment and photography, and to neutralize acids.

sodium hydrogen carbonate

Also known as bicarbonate of soda, a white, crystalline, mildly alkaline substance ($NaHCO_3$), manufactured by the ammonia-soda process. A solution of sodium hydrogen carbonate behaves as an alkali, neutralizing acids to form water, carbon dioxide and a salt. Sodium hydrogen carbonate is used in medical treatments as an antacid, in baking powders (baking soda) and dry-powder fire extinguishers.

sodium hydroxide

Also known as caustic soda, the commonest alkali (NaOH). Both the solid and the solution are extremely corrosive. It is prepared industrially from sodium chloride by the electrolysis of concentrated brine, where chlorine is the main product and sodium hydroxide is a byproduct. It is used to neutralize acids, in the manufacture of soap and paper, and in the treatment of effluents to remove heavy metals (as hydroxides).

sol

A **colloid** of very small particles, retaining the characteristics of a liquid.

solute

Any substance that is dissolved in another substance (**solvent**) to form a solution.

solution

A homogeneous mixture of two or more substances. One of the substances (generally a liquid) is the **solvent** and the others (**solutes**) are said to be dissolved in it. The constituents of a solution may be solid, liquid or gaseous. The solvent is normally the substance that is present in greatest quantity; however, if one of the constituents is a liquid, this is considered to be the solvent even if it is not the major component.

Solvay process

See **ammonia-soda process**.

solvent

Any substance, usually a liquid, that will dissolve another substance (**solute**) to form a solution. Solvents can be divided into polar solvents, such as water and liquid ammonia, and non-polar solvents, such as benzene and ethoxyethane, depending on their ability to dissolve differently bonded substances.

spectroscope

Any apparatus used in the study of the spectra associated with atoms or molecules. Spectroscopes can be used to identify unknown compounds, and are invaluable analytical tools in chemistry, astronomy, medicine and industry. Emission spectroscopes are used to analyze the characteristic series of sharp lines in the spectrum produced when an element is heated. Thus an unknown mixture can be analyzed for its component elements. Absorption spectroscopes are used to analyze atoms and molecules, which absorb energy in a characteristic way. More detailed structural information can be obtained using infrared spectroscopes (con-

cerned with molecular vibrations) or nuclear magnetic resonance (NMR) spectroscopes (concerned with interactions between adjacent atomic nuclei).

CONNECTIONS

TESTING FOR DRUGS **118**
SPECTROSCOPIC ANALYSIS **142**

spinneret

A disk perforated with fine holes of defined cross-section through which molten polymers are extruded in order to produce continuous filaments of synthetic fibers.

spontaneous combustion

The ignition of any chemical or material that is not initiated by the direct application of a flame. Sodium chlorate, for example, can react with its surroundings, usually by oxidation, to produce sufficient internal heat so that combustion occurs. Spontaneous combustion can occasionally occur in rags soaked with flammable liquids and in hay, when fermentation raises the temperature to burning point. *See* **exothermic reaction**.

starch

A carbohydrate macromolecule that consists of varying proportions of two glucose polymers (polysaccharides): straight-chain (amylose) and branched (amylopectin) molecules. It is widely produced by plants as a food store and is therefore a major energy source for animals. In its purified form, starch is a white powder. It is commercially used to stiffen textiles and paper, as a raw material for making various chemicals; and in the food industry as a thickening agent.

CONNECTIONS

ORGANIC CHEMISTRY **82**
CARBON, HYDROGEN AND OXYGEN **88**

stationary phase

See **chromatography**.

stereochemistry

The study of the different spatial arrangements (stereoisomers) of atoms in a molecule. In organic chemistry, the stereoisomers arise from the tetrahedral directions of the four different covalent bonds of carbon. Stereoisomerism can be divided into two types. In optical isomerism (see optical activity) two chemically identical structures with different configurations of their atoms can rotate plane-polarized light in different directions. In geometrical isomerism the

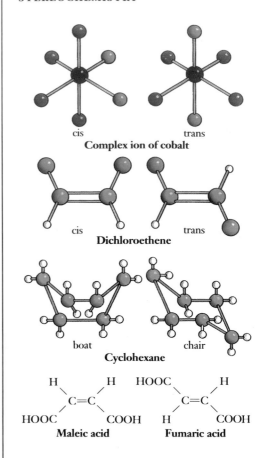

cis   trans
Complex ion of cobalt

cis   trans
Dichloroethene

boat   chair
Cyclohexane

H   H   HOOC   H
  \   /   \   /
   C=C   C=C
  /   \   /   \
HOOC   COOH   H   COOH
Maleic acid   **Fumaric acid**

double bond between two carbon atoms prevents the rotation of the carbons atoms, each of which has two different atoms or groups joined to it. This leads to two different isomers with *cis* and *trans* configurations. These geometrical isomers may have very similar chemical properties, but differing physical properties (for example, melting points).

stereoisomerism

See **stereochemistry**.

steroid

Any of group of lipids derived from a saturated compound called cyclopentano-perhydrophenanthrene, which has a complex molecular structure consisting of four carbon rings. One of the most important groups of steroids is the steroid alcohols (sterols). Other steroids include the sex hormones (such as testosterone and estrogen), the corticosteroid hormones (produced by the adrenal gland) and bile acids.

CONNECTIONS

CARBON, HYDROGEN AND OXYGEN **88**
MEDICAL DRUGS **114**

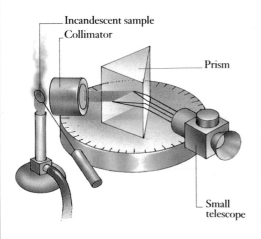

Incandescent sample
Collimator
Prism
Small telescope

styrene
See **phenylethene**.

subatomic particle
Any particle that is smaller than an atom. Subatomic particles may be indivisible elementary particles, such as the **electron** and quark, or they may be composites, such as the **proton**, **neutron** and alpha particle (which is equivalent to the nucleus of the helium atom and is a product of radioactive decay).

sublimation
The direct conversion of a solid to a vapor without passing through the liquid phase. A number of substances that do not sublime at atmospheric pressure can be made to do so by reducing the pressure. This is the principle of freeze-drying, during which ice sublimes at low pressure.

substrate
The compound or mixture of compounds acted on by an enzyme in a biochemical reaction.

sugar
Any of a group of colorless, sweet-tasting, soluble crystalline carbohydrates. Sugars are classified as monosaccharides, each molecule being composed of a single straight-chained or ring-shaped carbohydrate unit (for example, glucose and fructose); disaccharides, made up of two similar or different monosaccharide units (for example, sucrose and lactose); and polysaccharides, made up more than two carbohydrate units (for example, starch and cellulose).

> **CONNECTIONS**
>
> CARBON, HYDROGEN AND OXYGEN **88**
> NATURAL POLYMERS **96**

sulfate
Any salt or ester derived from sulfuric acid. Organic sulfates have the general formula R_2SO_4. Sulfate salts contain the ion SO_4^{2-}. Most sulfates are water-soluble (the exceptions are lead, calcium, strontium and barium sulfates), and require a very high temperature to decompose. The sulfate ion is detected in solution by using barium chloride or barium nitrate to precipitate the insoluble sulfate.

sulfide
Any compound of sulfur and another element in which sulfur is the more electronegative element. Compounds of sulfur with non-metals are covalent compounds (for example, the weak acid hydrogen sulfide). Metals form ionic sulfides based on the S_2^- ion and such compounds are salts of hydrogen sulfide. Sulfides occur naturally in a number of minerals and often have extremely unpleasant odors.

sulfur
A reactive, nonmetallic, yellow, brittle element, with the symbol S, the atomic number 16 and the relative atomic mass 32.06. It is found naturally and recovered from mineral deposits, using the **Frasch process,** and from natural gas and petroleum. It occurs in several allotropic forms, including plastic sulfur (*see* **allotrope**).

sulfur dioxide
A pungent gas or colorless liquid (SO_2), produced by burning sulfur in air or oxygen. The compound is a reducing agent, and is used widely for disinfecting food vessels and equipment, and as a preservative in some food products. It occurs in industrial flue gases and is a major cause of atmospheric pollution, notably **acid rain** (forming a mixture of sulfuric and sulfurous acids when mixed with water).

sulfuric acid
A dense, viscous, colorless liquid (H_2SO_4) that is extremely corrosive. It gives out heat when added to water and can cause severe burns. Sulfuric acid is used extensively in the chemical industry, in the refining of gasoline, and in the manufacture of fertilizers, detergents, explosives and dyes. It forms the acid component of automobile batteries. Its traditional name is oil of vitriol.

> **CONNECTIONS**
>
> ACIDS, BASES AND SALTS **62**
> MAKING ACIDS AND BASES **76**

superglue
The general name for fast-setting, powerful adhesives formed by polymerization.

surface coating
Treatment of the surface of any material to improve properties such as wear resistance, electrical conductivity or the ability to accept a dye. *See also* **anodizing**.

surfactant
Any substance added to a liquid to increase its wetting or spreading properties (that is, increase its surface tension). **Detergents** or **soaps** dissolved in water are common examples. The name is derived from a contraction of surface-active agent.

synthesis
The formation of chemical compounds from more elementary compounds. The synthesis of a drug, for example, can involve several stages from the initial material to the final product; the complexity of these stages is a major factor in the cost of production.

> **CONNECTIONS**
>
> SYNTHETIC POLYMERS **98**
> NATURAL DRUGS **116**

synthetic drug
Drugs that are synthesized in the laboratory or by an industrial process and that do not occur naturally. Such synthetic drugs are often based on more active modifications of natural drugs. Most pharmaceutical drugs are now made synthetically. *See* **chemotherapy** and **molecular modeling**.

synthetic rubber
Any of a number of artificial, rubberlike compounds that have better properties than natural rubber in terms of, for example, oil resistance and oxidation. Synthetic rubbers are usually polymers of isoprene or its derivatives, such as Neoprene, or copolymers of vinyl chloride and vinyl acetate.

systematic nomenclature
The most common modern system for naming chemicals that gives useful information about the structure of a substance. For organic compounds, the systematic name is derived from a suitable alkane; for example, ethanoic acid, CH_3COOH, and propanone, CH_3COCH_3 (traditionally known as acetic acid and acetone) are named after their "parent" alkanes, ethane (CH_3CH_3), and propane ($CH_3CH_2CH_3$). For inorganic compounds, the oxidation number of the metal is given; for example, iron(III) oxide, indicating Fe_2O_3. *See also* **chemical name**.

table salt
Another name for sodium chloride (NaCl), a white crystalline solid, found dissolved in sea water and as rock salt in large deposits and salt domes. This common salt is used extensively in the food industry as a preservative and for flavoring, and in the chemical industry in the making of chlorine, hydrochloric acid and sodium.

> **CONNECTIONS**
>
> MIXTURES AND COMPOUNDS **52**
> TYPES OF BONDS **54**
> BONDS AND STRUCTURES **56**

tautomerism

A state of dynamic equilibrium between two spontaneously convertible isomers (tautomers). The conversion between the two isomers is brought about by the change in position of a hydrogen atom in the molecular structure. The two tautomers are known as the "keto" and "enol" forms.

temperature

The state of hotness or coldness of an object. Temperature is measured on the Celsius, Kelvin or Fahrenheit scales.

test paper

A piece of paper coated with an indicator reagent which changes color when testing particular substances. The test paper can thus be used to identify certain specific properties of the substance tested. In the case of litmus paper, used to test **pH**, blue indicates the presence of an alkali and red indicates the presence of an acid.

thermochemistry

The study of the changes of heat energy that occur during a chemical reaction. Practical thermochemistry mainly uses calorimetry, to find standard heats of reaction.

thermoplastic

Plastics that always soften when warmed and then harden again as they cool. Thermo-softening plastics include polyethene, polystyrene, nylon and polyester. They can be recycled if collected in a clean state.

thermosetting plastics

Plastics that remain rigid once set and do not soften when warmed. Thermosetting plastics include bakelite, epoxy resins (used in paints, varnishes, and laminates, and as adhesives) and polyurethane. They cannot be recycled.

TAUTOMERISM

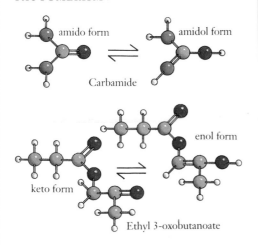

amido form   amidol form

Carbamide

enol form

keto form

Ethyl 3-oxobutanoate

thiosulfate

Any salt derived from thiosulfuric acid and containing the $S_2O_3^{2-}$ ion. Under acid conditions thiosulfates decompose to give sulfur and the hydrogen sulfite ion (HSO_3^-).

titanium oxide

A white, naturally-occurring oxide of titanium (TiO_2). It is found in many mineral forms, particularly rutile, anastase and brookite, and is used as a white pigment and as a filler for plastics and rubber.

titration

An analytical technique to determine the concentration of one compound in a solution by determining how much of it will react with another compound of known concentration. A measured quantity of one of the solutions is placed into a suitable reaction vessel. The other solution is added a little at a time from a graduated delivery device (burette). The completion of the chemical reaction (the end point) is determined by the change in color of an **indicator** or by using an electrochemical detector. Common titrations are between acids and alkalis.

transition element

Any of a group of metals on the **Periodic Table** of the element, that have incomplete inner electron shells and exhibit variable valence. Common transition metals include cobalt, copper, iron and molybdenum. Most are excellent conductors of heat and electricity, generally form colored compounds, and have high melting and boiling points.

transuranium element

Any of the elements above uranium in the **Periodic Table**; that is, with an atomic number of 93 or higher. These elements possess heavier and more complex nuclei than uranium, and are all radioactive. Neptunium and plutonium are found in nature; all of the others can be produced from uranium by bombarding it with neutrons.

triglyceride

The term given to a fatty acid ester of glycerol (propan–1,2,3-triol) in which all three hydroxyl groups are substituted. In simple triglycerides all three fatty esters are identical; in mixed triglycerides two or three different fatty acid esters are present. Triglycerides are the major components of fats and oils and an important food store in living animals.

triple bond

Three covalent bonds between adjacent atoms, as in the alkynes (–C≡C–). *See* **covalent bond** and **double bond**.

universal indicator

A mixture of acid-base **indicators,** used to gauge the pH of a solution. Each component indicator changes color at a different pH value and so the mixture is capable of displaying a range of colors, according to the pH of the solution being tested, from red (at pH 0) to purple (at pH 14). The pH may be found by adding a few drops of the indicator or by using an absorbent **test paper** that has been impregnated with the indicator and noting any color change.

unsaturated compound

Any compound that has double or triple covalent bonds between the atoms of its molecules. Examples are alkenes and alkynes, in which the two adjacent atoms are both carbon, and ketones, in which the unsaturation exists between atoms of different elements. Unsaturated compounds can undergo addition reactions as well as substitution reactions. The laboratory test for unsaturated compounds is to add bromine water; if the test substance is unsaturated, the bromine water will be decolorized. The term unsaturated is also applied to a solution in which more solute may be dissolved. *See also* **saturated compound**.

unsaturated fat

A fatty acid that contains double bonds in the hydrocarbon chain. *See* **unsaturated compound**.

urea

A white crystalline solid ($CO(NH_2)_2$, the end product of many protein breakdowns in mammals, and synthesized by heating ammonia and carbon dioxide.

urea–formaldehyde resins

A group of synthetic polymer resins produced by the copolymerization of urea and formaldehyde (methanal) when heated. The resulting **thermosetting plastics** are translucent and can be easily colored by dyes and tints. They are nonflammable and resistant to weathering and many chemical agents. They are also used as adhesives.

CONNECTIONS

SYNTHETIC POLYMERS **98**

TYPES OF PLASTIC **100**

SHAPING PLASTICS **102**

---

urine tests

The biochemical analysis of the composition of urine. Such tests can be used to detect the presence of glucose or of drugs in the body, or to determine pregnancy.

VALENCE

Metals and cations	Non-metals and anions
VALENCE 1	
Ammonium NH_4^+	Chlorate (V) ClO_3^-
Copper (I) Cu^+	Chlorine Cl
Mercury (I) Hg^+	Hydrogen carbonate HCO_3^-
Silver Ag^+	Hydroxide OH^-
Sodium Na^+	Nitrate NO_3^-
VALENCE 2	
Barium Ba^{2+}	Carbonate CO_3^{2-}
Copper (II) Cu^{2+}	Oxygen O
Lead (II) Pb^{2+}	Sulfate (VI) SO_4^{2-}
Mercury (II) Hg^{2+}	Sulfite (IV) SO_3^{2-}
Zinc Zn^{2+}	
VALENCE 3	
Aluminum Al^{3+}	Nitrogen N
Chromium (III) Cr^{3+}	Phosphorus (III) P
Iron (III) Fe^{3+}	Phosphate (V) PO_4^{3-}
VALENCE 4	
Lead (IV) Pb^{4+}	Carbon C
Manganese Mn^{4+}	Silicon Si
Tin (IV) Sn^{4+}	Sulfur (IV) S

valence

The combining capacity of an atom or radical, determined by the number of electrons that it adds, loses or shares when it reacts with another atom. Valence is often indicated by a Roman numeral after the chemical name – sulfur (IV) – and varies according to whether the substance occures in its elemental form or in a compound with other elements; sulfur can have a valence of 2, 4 or 6. Elements that lose electrons (such as hydrogen and the metals) have a positive valence; those that add electrons (such as oxygen and other nonmetals) have a negative valence. In ionic compounds, the valence of an element is equal to the ionic charge; in sodium sulfide, Na_2S, sodium has a valence of 1 (Na^+) and sulfur has a valence of 2 (S^{2-}). In covalent compounds, the valence (covalence) is equal to the number of bonds formed; in carbon dioxide (CO_2) carbon has a valence of 4 and oxygen has a valence of 2.

van der Waals' force

A weak force of attraction between atoms and molecules, derived from the movement of electrons within atoms. It is 10 to 20 times weaker than the attractive force between atoms in covalent bonds.

vapor

A state of matter similar to a gas in which the molecules of a substance move in a random fashion and are spaced far apart. The distance between the molecules, and therefore the volume of the vapor, is limited only by the walls of the containing vessel. A vapor can be liquefied by increased pressure.

vaporization

The change of a liquid directly to a vapor at a temperature below its boiling point. Volatile liquids vaporize at room temperature.

vitamin

Any of a number of organic compounds that are necessary in small quantities for normal body function. There are about 14 vitamins, which are divided into water-soluble vitamins, such as vitamins B and C, and fat-soluble vitamins, such as vitamins A, D and K. Many act as coenzymes, small molecules that enable enzymes to function effectively. Although animals are unable to manufacture vitamins themselves, they are normally present in adequate amounts in a balanced diet.

volumetric analysis

A quantitative analysis procedure for measuring volumes. For gases, the main technique involves reacting or absorbing gases in graduated containers over mercury and measuring any changes in volume. For liquids, the main technique is **titration**.

vulcanization

The process of treating rubber with sulfur or sulfur-containing compounds to improve its physical properties, particularly its resistance to wear. The sulfur is absorbed by the rubber, and this can be achieved by either heating the raw rubber with sulfur at high temperatures (135–160°C) or by treating cold rubber sheets with a solution of S_2Cl_2.

water

A liquid oxide of hydrogen (H_2O). It begins to freeze at 0°C and to boil at 100°C. When liquid, it is virtually incompressible; frozen, it expands by 1/11 of its volume. At 4°C, one cubic centimeter of water has a mass of one gram; this is its maximum density, forming the unit of specific gravity. It has the highest known specific heat and acts as an efficient solvent, particularly when heated.

wax

Any of a number of solid or semi-solid fatty substances composed of esters, fatty acids, free alcohols and solid **hydrocarbons**. Mineral waxes are obtained from petroleum and vary in hardness from soft petroleum jelly to hard paraffin wax, commonly used in candles. Animal waxes include beeswax and lanolin – often added to cosmetics as a softening agent. Vegetable waxes usually occur as a waterproof coating on plants.

CONNECTIONS

MAKING HYDROCARBONS 92
COSMETIC CHEMISTRY 126

wettability

The extent to which a solid can be wetted by a liquid, measured by the force of adhesion between the solid and liquid phases. Wetting agents lower the surface tension of a liquid so that it can more effectively spread across or penetrate a solid. *See* **soap** and **detergent**.

X-ray crystallography

The study of the atomic and molecular structure of crystalline substances by passing X rays through them. A sample of a substance is ground to a fine powder and exposed to X rays delivered from various angles. The diffraction pattern (a pattern of spots) produced by the diffracted X rays on a photographic plate can then be compared with reference standards for identification.

zeolite

Any of a group of naturally-occurring or synthetic hydrated aluminum silicates, also containing sodium, calcium, barium, strontium or potassium, in which water molecules are held in cavities in the crystal lattice. Zeolites are used as **molecular sieves** to separate mixtures because they absorb selectively. They have a high ion-exchange capacity (which allows them to soften hard water), and can be used to make gasoline, benzene and toluene from low-grade raw materials such as coal and methanol.

X-RAY CRYSTALLOGRAPHY

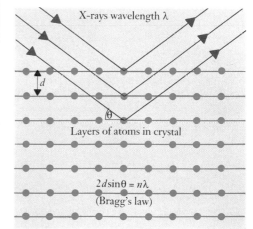

X-rays wavelength λ

d

θ

Layers of atoms in crystal

$2d\sin\theta = n\lambda$
(Bragg's law)

1

ATOMS
and Molecules

CHEMISTS have always worked with atoms. The study of
what happens between atoms during chemical changes is
the basis of their science. Yet it was not until the invention
of the scanning tunneling microscope in the 1980s that chemists
were actually able to see atoms for the first time.

The ancient Greeks believed that atoms were particles that
cannot be divided again, and named them "atomos" (indivisible).
We now know that atoms are made up of even smaller particles:
protons, which carry a positive electrical charge; electrons, which
have a negative electrical charge; and neutrons, which are neutral.
The center, or nucleus, of most atoms is made up of neutrons and
protons, and carries a positive charge. Only the hydrogen atom has a
nucleus which consists of just one proton. The electrons move
around the nucleus in a series of "shells", or orbitals, held in place by
the attraction of the positive charge of the protons in the nucleus.

Everything around us is made up of combinations of chemical
elements – substances that consist of only one type of atom, and
cannot be broken down into a simpler substance. Because atoms are
the smallest unit of a chemical element that retain the characteristics
of the element, they are the building blocks of all things – and
chemistry touches on all aspects of our lives.

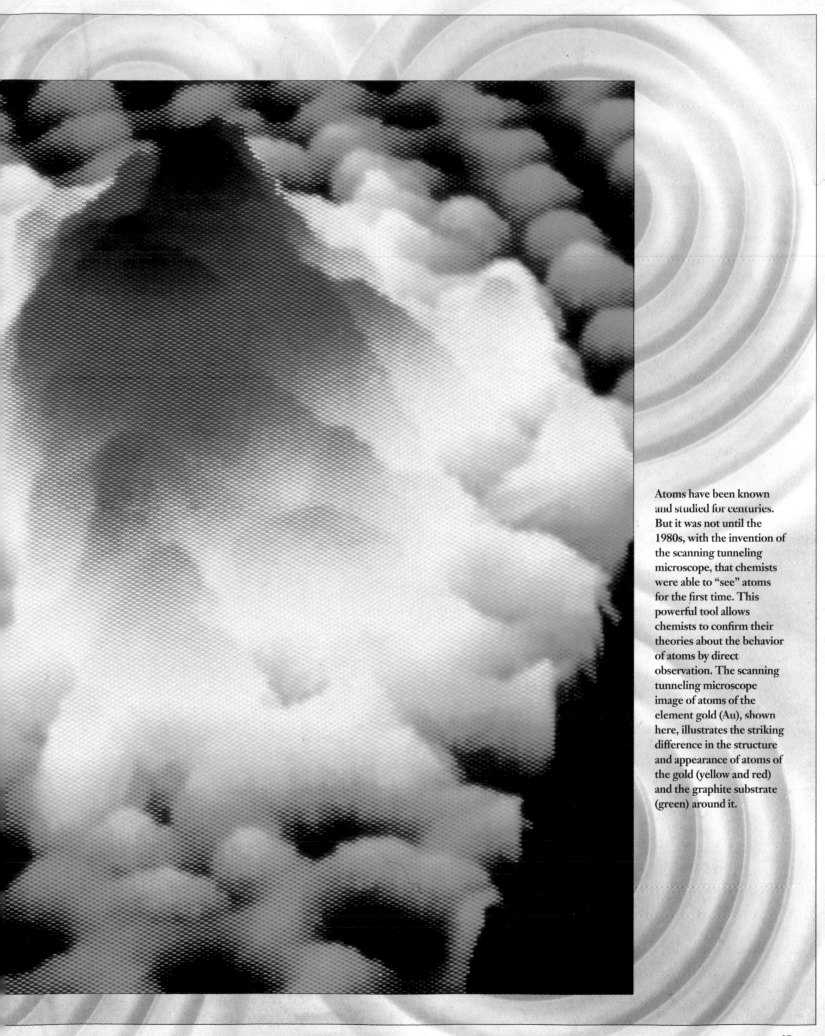

Atoms have been known and studied for centuries. But it was not until the 1980s, with the invention of the scanning tunneling microscope, that chemists were able to "see" atoms for the first time. This powerful tool allows chemists to confirm their theories about the behavior of atoms by direct observation. The scanning tunneling microscope image of atoms of the element gold (Au), shown here, illustrates the striking difference in the structure and appearance of atoms of the gold (yellow and red) and the graphite substrate (green) around it.

THE ELEMENTS

I**T MIGHT** seem impossible to predict the chemical and physical properties of an element without carrying out experiments. But because of the way that atoms interact with each other, it is only necessary to know the number of protons in the nucleus of the atom of that element – referred to as the atomic number – to predict quite a lot about an element's physical and chemical properties. The great predictive tool is a chart known as the Periodic Table.

In the Periodic Table, elements are arranged in order of increasing atomic number. Elements with similar physical and chemical properties are found at definite intervals, or periods, of atomic number. This periodicity makes it possible to predict the characteristics of an element simply by knowing its position in the Periodic Table. When the table was first compiled, periodicity made it possible for chemists to predict the existence of elements such as germanium (Ge), which were not then known. Now it allows chemists to predict the properties of superheavy elements that have not yet been synthesized.

Metals make up most of the elements and take up all of the left-hand side of the Periodic Table. The nonmetals appear on the right, separated from the metals by the metalloids. In the Periodic Table,

▷ Although rare on Earth, Helium (He) is common in the Universe. It was first detected in the Sun's spectrum and named after *helios*, the Greek word for the Sun. Helium, a gas, is lighter than air. Because its outer electron shell is full, it does not seek electrons and rarely reacts with other elements.

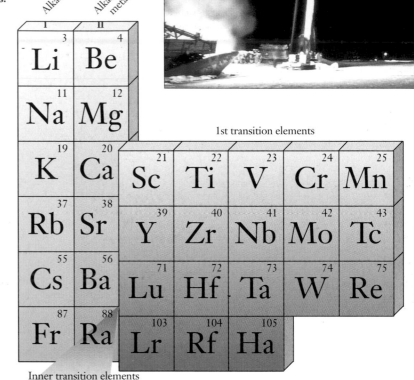

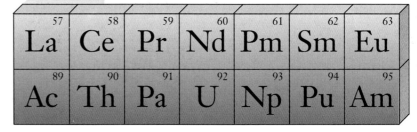

◁ **Magnesium (Mg), a reactive silvery-white metal, burns readily in air with a dazzling white flame. It is often used in flares, flashbulbs and other pyrotechnics. Magnesium is the eighth most common element, and is found in rocks such as dolomite and brucite. It is light but strong, and forms useful alloys with aluminum and other metals.**

elements are arranged in vertical columns, known as groups, and in horizontal rows known as periods. Elements in the same group generally have similar chemical properties.

The Group I elements, or alkali metals, react vigorously in water to create strong alkaline solutions. Group II elements, the alkaline earth metals, never occur naturally in their pure form, but instead occur as compounds in minerals which form rocks. The transition metals, which are usually divided into three blocks and occur between groups II and III, are hard, tough and shiny. Aluminum (Al), which is the third

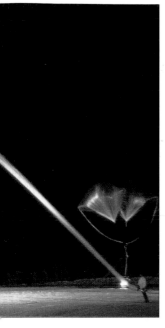

■ The Periodic Table groups elements according to their physical and chemical properties. An element's place is determined by its number of protons: hydrogen (II) has one, helium (He) two; and so on. The first table was published in 1869 by the Russian chemist Dmitri Mendeleyev. The horizontal rows in the Table are known as periods. From left to right, each element has one more electron in its outer shell than the last. Vertical columns on the Table are known as groups. Elements in the same group all have the same number of electrons in their outer shell, and, as a result, they tend to have similar chemical properties.

So far 105 elements have been discovered, named and listed on the Periodic Table. It is possible that up to 15 more may have existed previously, when the Earth was formed, and these may be rediscovered through synthesis in high-energy particle accelerators.

The atomic number of each element appears at the top of the element's square in the Periodic Table. Isotopes are atoms of the same element that have different numbers of neutrons in their nuclei, although the number of protons remains the same in the isotope as in its original element. Some elements have radioactive isotopes: the nuclei spontaneously disintegrate, sending out high-energy particles as they do so.

The total of neutrons and protons in an atom is known as its atomic mass. For convenience, chemists often refer to the relative atomic mass of an element (formerly known as atomic weight). This is found by comparing the mass of an "average" atom of an element, taking into account the proportions of all its isotopes, with a reference value equivalent to one-twelfth of the mass of an atom of an isotope of carbon, carbon-12.

		Halogens	Hydrogen	Noble gases
			1 H	2 He

Boron group III	Carbon group IV	Nitrogen group V	Oxygen group VI	VII	VIII
5 B	6 C	7 N	8 O	9 F	10 Ne
13 Al	14 Si	15 P	16 S	17 Cl	18 Ar
31 Ga	32 Ge	33 As	34 Se	35 Br	36 Kr
49 In	50 Sn	51 Sb	52 Te	53 I	54 Xe
81 Tl	82 Pb	83 Bi	84 Po	85 At	86 Rn

2nd transition elements   3rd transition elements

26 Fe	27 Co	28 Ni	29 Cu	30 Zn
44 Ru	45 Rh	46 Pd	47 Ag	48 Cd
76 Os	77 Ir	78 Pt	79 Au	80 Hg

64 Gd	65 Tb	66 Dy	67 Ho	68 Er	69 Tm	70 Yb	Lanthanides
96 Cm	97 Bk	98 Cf	99 Es	100 Fm	101 Md	102 No	Actinides

▽ Lead (Pb), a soft, heavy metal, weighs down a diver's boots. Its symbol derives from the Latin word for lead (*plumbum*).

▽ Sulfur (S) has been mined since ancient times; today most sulfur is used for making sulfuric acid, a key industrial chemical.

most abundant element on Earth, occurs in Group III. Carbon (C) and silicon (Si), the second most abundant element, are included in Group IV. Group V elements include nitrogen (N) and phosphorus (P). Oxygen (O), the most abundant element on Earth, is included in Group VI, along with sulfur (S). The Group VII elements, or halogens, are nonmetals and are so reactive that they are usually found combined with other elements in a salt. The Group VIII elements, the noble or inert gases, are very different. Their outer electron shells are full, which makes them almost totally nonreactive.

MIXTURES AND COMPOUNDS

When two or more atoms bond together, they form a molecule. Molecules can be made up of several atoms of the same element, or of atoms of different elements. They can come together in many ways to produce mixtures, solutions, emulsions and even new compounds; all these are important to the chemist, although compounds are what people most often think of as the special province of the chemist.

A common everyday example of a mixture can be found in a typical trash can. It probably contains a variety of different substances such as glass, paper, plastic and metal all mixed together. In the trash can no chemical reactions have taken place and the different components of the waste are mixed up but not bonded together chemically. It is an easy – if unpleasant – task to separate physically the different components using characteristics such as size, texture, color and density.

A cup of coffee or tea contains examples of both compounds and solutions. Compounds are molecules that contain atoms from at least two different elements, and water (H_2O) is the commonest and most abundant compound on Earth. In water, as in all compounds, the atoms are held together by chemical bonds. There are several kinds of chemical bond, but all involve the sharing of electrons between the atoms of the molecule. For such a bond to form or be broken requires a chemical reaction to take place, and the individual elements cannot be joined or separated by physical means such as shaking, pressing or filtering. The molecular structure of water makes it an excellent solvent, and water is often known as the universal solvent because so many substances can be dissolved in it. The solvent action of water is essential for making a hot drink such as coffee or tea. Hot water is used as a solvent to dissolve the flavorful juices from the ground coffee beans or the tea leaves. Some people also dissolve a solid solute (sugar) and a liquid solute (milk or cream). These solutes are readily soluble in water. In the drink, they are broken down into molecules or ions which become evenly dispersed throughout the solvent. The solute molecules eventually disperse through the hot drink without any help, but they dissolve more quickly if the drink is stirred. As well as the dissolved molecules, a cup of coffee contains tiny particles of the ground beans held

KEYWORDS

BOND
COMPOUND
EMULSIFIER
EMULSION
MIXTURE
MOLECULE
SOLUTE
SOLUTION
SOLVENT

in suspension, but these are not chemically altered as are the solutes. In hard water areas where the water contains calcium, magnesium and bicarbonate ions, a film or scum may appear on the surface of the drink; this can be avoided by adding a weak acid such as lemon juice instead of milk. The scum is composed mainly of calcium carbonate crystals, which dissolve readily in acid.

Margarine is a common example of an emulsion – a mixture of two liquids that do not dissolve in each other. In an emulsion, the components are not broken down and tiny droplets of one liquid are suspended in the second liquid. Margarine is an emulsion of fats, oils and milk, which is mainly water. Normally oil and water do not mix, because the differences in their molecular structure mean that they have a stronger attraction for their own kind of molecule than for each other. It is possible to overcome this problem by shaking the mixture vigorously. Thus when making a salad dressing the oil can be emulsified with vinegar, which is also mostly water. The shaking breaks up the two liquids into tiny droplets and causes them to form a temporary emulsion. When the droplets join together again, the two liquids soon separate, with the oil floating on the vinegar.

In margarine a more permanent emulsion is required, so an emulsifier is used. Emulsifiers are molecules that contain one end which is attracted to oil and one end which is attracted to water. The natural emulsifier lecithin, which is found in egg yolks, is often used for this process.

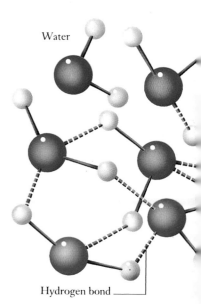

1 These piles of iron filings and yellow sulfur powder are examples of elements. They cannot be broken down into simpler substances. When iron and sulfur are merely mixed together 2, no chemical reaction occurs. The two components of the mixture can be separated quite easily by using a magnet to remove the iron filings 3. But if the iron/sulfur mixture is heated, a chemical reaction does occur, and the two components can no longer be readily separated 4. The compound which results is known as iron sulfide (FeS). In nature, FeS occurs as the mineral pyrite, also known as "fool's gold" because its golden color deceived some gullible prospectors.

SALT AND WATER

Sodium chloride (NaCl) or common salt is a compound made up of sodium and chlorine atoms. Because the sodium atoms have a positive charge and chlorine atoms have a negative charge, they are ions, held together by ionic bonds. NaCl dissolves readily in water. When salt dissolves in water it splits into its component ions, which become surrounded by water molecules. The negatively charged Cl^- and positively charged Na^+ ions are attracted to different ends of the water molecules. When the water is removed, the Na and Cl recombine to form salt crystals. Water is a very common but unusual compound held together by a weak hydrogen bond. Although made up of two gases, hydrogen and oxygen, water at room temperature is a liquid rather than a gas.

Water

Hydrogen bond

◁ Ice cream is a familiar example of an oil-in-water emulsion. In commercially made ice creams, especially soft ice creams, the fat and water are held together by emulsifiers such as glyceryl monostearate. In contrast, sand is a mixture, and sea water contains salt and other minerals in solution.

▷ Although the chemical reaction between iron and sulfur needs a supply of heat to make it begin, a lot of heat is given off as the vigorous reaction proceeds. It is an exothermic reaction.

2

3

4

Water molecules

Hydrated chloride ion

Sodium chloride crystal

Hydrated sodium ion

Chloride ion

Sodium ion

TYPES OF BOND

ALTHOUGH electrons are the smallest atomic particles – they are more than 1800 times smaller than a neutron or proton – they determine the chemical properties of atoms. Electrons are largely responsible for the way that atoms react, or bond, with other atoms. Atoms join together to form a molecule when the pull of the nucleus of a nearby atom is stronger than the pull of their own nucleus, and the electron moves to the outer shell of the neighboring atom, or becomes shared between the atoms. Understanding the role of electrons in bonding provides the key to understanding chemical reactions.

Atoms are always seeking chemical stability. Those that have an incomplete outer shell of electrons try to join up chemically with other atoms. In contrast, elements that have a full outer shell of electrons are very stable. (An example is neon, one of the "noble gases.")

Atoms join together to form molecules and compounds by means of chemical bonds. A chemical reaction, which involves the making or breaking of bonds, occurs spontaneously only if the products are more stable than the atoms that react. In bonding, atoms either lose, gain or share electrons, and the number of electrons that an atom has available to participate in the bonding is known as its valence, or combining power.

Atoms can combine in many ways. As a result, two compounds with the same chemical composition may have different forms, or isomers, which can also have different chemical

CHEMICAL BONDING

There are three main types of bond: ionic, metallic and covalent. When an atom loses or gains electrons it becomes electrically charged, and is known as an ion. An ionic bond is formed when ions with opposite charges are held together by electrical attraction, and form a regular array, or crystal, also known as an ionic lattice. In common salt (NaCl), sodium atoms lose an electron and take on a positive charge, while the chlorine atoms gain an electron to take on a negative charge.

In metallic bonding, a lattice is formed when all the metal atoms share their outer electrons to form a sea of electrons. Metallic bonds are very strong, and thus metals tend to have high melting and boiling points. In the metallic lattice the electrons can move freely and this explains why metals are such good conductors of electricity and heat.

In covalent bonding, atoms do not gain or lose electrons; instead, they share their electrons with another atom. When only one electron is shared between the atoms, as for example in the fluorine molecule (F_2), a single covalent bond is formed. When two electrons are shared, as in carbon dioxide (CO_2), the resulting covalent bond is known as a double bond. In acetylene (ethyne) (C_2H_2), where the two carbon atoms are held together by a triple bond, three electrons are shared.

Electrons in covalent bonds are not always shared equally between the bonded elements. Where the sharing is unequal, as in a molecule of water (H_2O), the bond is known as a polar bond, and the resulting molecule has both a negatively and a positively charged end.

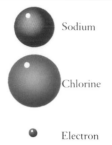

Sodium

Chlorine

Electron

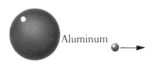

Aluminum

Electron

Fluorine

Oxygen

Hydrogen

Electron

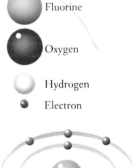

Fluorine atom

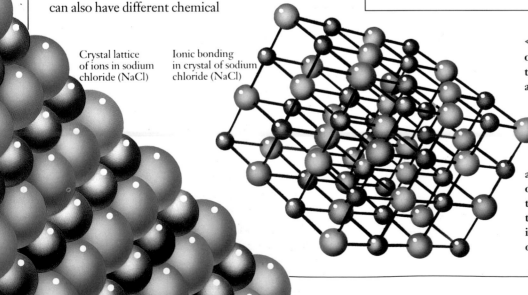

Crystal lattice of ions in sodium chloride (NaCl)

Ionic bonding in crystal of sodium chloride (NaCl)

◁ Common salt, NaCl, is one of many compounds that form crystals. Crystals are bounded by definite faces which intersect at characteristic angles. The shape of the crystal RIGHT reflects the three-dimensional arrangement of the atoms or molecules that make up the crystal, and the study of the arrangement of atoms in a crystal can shed light on its chemical properties.

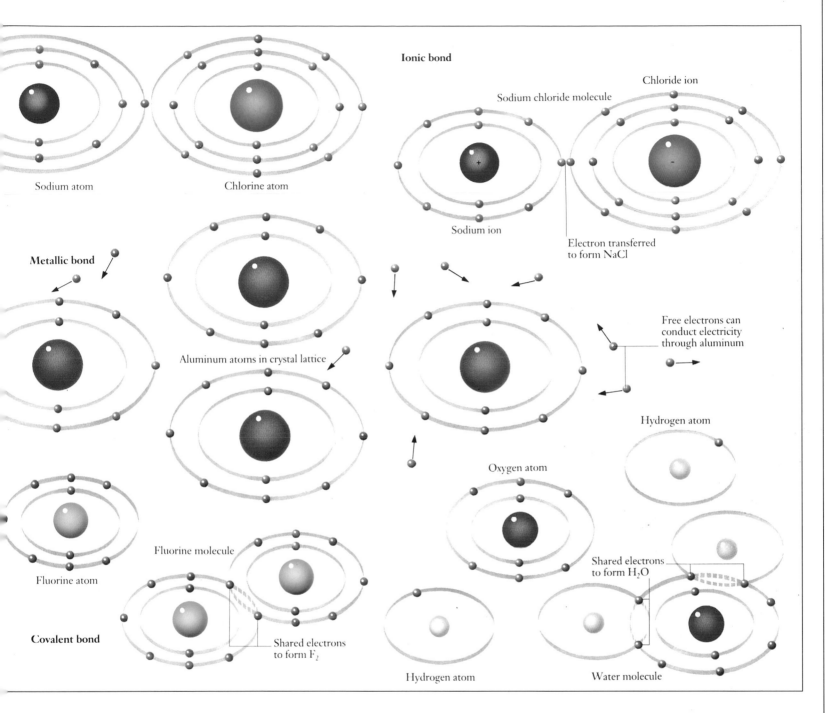

Ionic bond

Sodium atom

Chlorine atom

Sodium chloride molecule

Chloride ion

Sodium ion

Electron transferred to form NaCl

Metallic bond

Aluminum atoms in crystal lattice

Free electrons can conduct electricity through aluminum

Hydrogen atom

Oxygen atom

Fluorine molecule

Fluorine atom

Shared electrons to form H_2O

Covalent bond

Shared electrons to form F_2

Hydrogen atom

Water molecule

◁ **Like common salt (NaCl), the mineral pyrite (FeS) BELOW LEFT is made up of just two elements. And like the sodium and chlorine in salt, the iron and sulfur atoms are arranged in a face-centered cubic lattice. In contrast to salt crystals, which are usually cube-shaped, pyrite crystals can be cubic, octahedral, or have 12 pentagonal facies in a shape known as pyritohedral.**

properties. Understanding the properties of isomers is crucial in oil refining. To produce an automobile fuel with the optimum volatility (to make the car easy to start from cold) and the right octane number (so the car runs smoothly and with enough power), refiners must achieve the right balance of straight-chain and branched isomers of alkanes, the chief components in crude oil. Although both isomers have the same chemical composition, their molecular shapes are different, and as a result they perform differently as a fuel. A fuel with too many straight-chain alkanes would be too volatile, and a fuel without enough branched alkanes would have a very low octane rating.

BONDS AND STRUCTURES

WHEN two oxygen atoms join with one carbon atom by means of double covalent bonds, they form the relatively harmless gas carbon dioxide (CO_2). But when the same elements are bonded together in carbon monoxide (CO), the result is very different: lone pair electrons allow the carbon monoxide to bond as a ligand to other substances, and the compound is a poisonous gas that can kill.

A chemical formula alone is not a reliable indicator of how a substance looks or what its physical properties are. These properties are greatly influenced by the way in which its molecules are attracted to one another. The stronger the attractive bonds, the higher the melting and boiling point of the substance.

The properties of the silicon bouncing putty sold as a toy for children are also the result of a combination of attractions between molecules and bonds within the molecule. Here, boron atoms, which easily form coordinate bonds with neighboring chains of atoms, are used to replace some of the silicon in the silicone polymer chains. When the putty is pulled smoothly, the silicone chains slide over each other as the boron atoms form coordinate bonds with successive oxygen atoms in the neighboring chain; this causes the putty to stretch. But when it is pulled sharply, it breaks, because the covalent bonds between the boron atoms in the chains are broken.

The forces between molecules also determine the physical properties of substances. Some molecules, such as water, are held together by polar covalent bonds, in which the electrons are shared unequally. This results in a slight positive electrical charge at one end and a slight negative charge at the other. These opposite charges attract, and dipole-dipole attraction

KEYWORDS

ALLOTROPE

BOND

CARBON DIOXIDE

COVALENT BOND

DOUBLE BOND

HYDROGEN BOND

TRIPLE BOND

VAN DER WAALS' FORCE

WATER

▷ Carbon illustrates the wide variation in properties that can result from different arrangements of atoms. It exists naturally in four forms, or allotropes: amorphous carbon, graphite, diamond and buckminsterfullerene RIGHT. The variety of carbon compounds results from the fact that carbon has a valence of 4 and can form single, double and triple bonds to itself. It can also form four covalent bonds to other atoms. Amorphous carbon occurs in the soot formed during incomplete burning of hydrocarbon fuels, and consists of shapeless particles of carbon. In a similar way wood or soft coal can be incompletely burned to produce charcoal. In graphite, the carbon is arranged in thin sheets composed of a network of six-sided carbon rings. The sheets are held together by weak van der Waals' forces.

Buckminsterfullerene

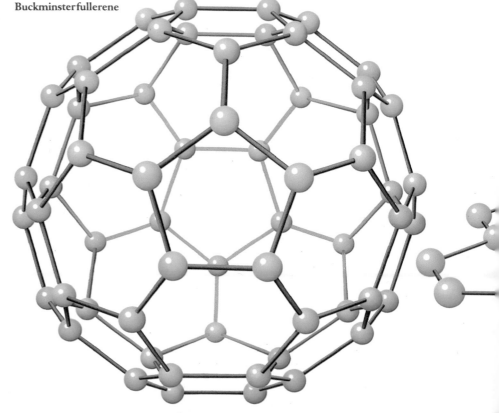

▷ Buckminsterfullerene (C_{60}), the fourth allotrope of carbon, was extracted and characterized from soot only in 1985. Its basic structure consists of five- and six-sided carbon rings arranged in a shape resembling a soccer ball or a geodesic dome (invented by the architect Buckminster Fuller; hence the name). This unique structure offers many possibilities for "doping" the molecule to change its electrical and magnetic characteristics by placing other compounds inside its hollow center. This has led to a whole new class of molecules known as fullerenes. Some doped forms have superconducting properties at relatively high temperatures. Others show useful ferromagnetic properties.

opposite charges attract, and dipole-dipole attraction pulls the molecules together. Hydrogen bonds are a very strong form of dipole dipole attraction. Although they are weak compared with normal chemical bonds, they have some significant effects. Hydrogen bonding in water is responsible for its high surface tension. This is why some insects can "float" on water. Hydrogen bonding also accounts for the relatively high boiling point of water. Energy (heat) is needed to break the bonds and separate the liquid molecules into the free individual molecules found in a vapor.

Intermolecular forces do not have to be strong to be powerful. Van der Waals' forces, the "flickerings" of positive and negative charges that occur on the surface of molecules due to the constantly shifting position of the electrons, pull nonpolar molecules together if there is no other stronger force to get in the way. In graphite, covalent bonds hold together the carbon sheets themselves, but van der Waals' forces hold the layers together. The weak van der Waals' forces allow the layers to slide over each other. This is why graphite is such a good lubricant.

■ Diamond BELOW, a form of carbon, is one of the hardest natural substances. Its value as a gemstone depends on the way it is cut, but its main use is industrial rather than decorative: its hardness makes it an excellent cutting tool. The mineral graphite BELOW LEFT, by contrast, is one of the softest forms of carbon. It makes a good lubricant and electrical conductor.

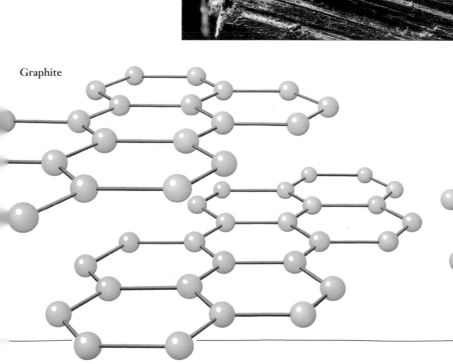

Graphite

Diamond

◁ In graphite CENTER LEFT, a network of six-sided carbon rings forms thin sheets held loosely together. In diamonds LEFT, each carbon atom is covalently bonded to four others to form a tightly packed three-dimensional lattice of tetrahedrons.

CHEMICAL *Reactions*

THE ART OF THE CHEMIST LIES in understanding and controlling the chemistry that causes chemical reactions. Chemical reactions may take place either explosively or gradually, and some may not take place at all without help.

It is the type of reactant that determines the rate at which the reaction will take place, or whether it will occur at all. In general terms, atoms seek stability by having a complete outer shell of electrons. An element which has just one electron in its outer shell, such as hydrogen (which has just one proton and one electron in total), easily donates that electron. Hydrogen forms many compounds, ranging from water, acids and bases, to hydrocarbons and other organic molecules. Metals and halogens such as sodium and chlorine are so reactive that they can be difficult to control.

An understanding of how chemical bonds are broken and formed helps chemists to initiate reactions, handle reactive elements safely, and control the rates of reactions. Increasing temperature, pressure or concentration can provide the activation energy needed to start off a reaction and increase its rate. Catalysts – substances that speed up the rate of reaction without undergoing any permanent chemical change themselves – can make even the least reactive elements join together to form compounds.

Every industrial chemical process, no matter how small, begins on a laboratory bench. The chemical laboratory is where new chemical ideas are born. After a potentially useful reaction has been discovered, it can be scaled up for use in an industrial setting. Developing an understanding of how chemical reactions work is an important first step toward creating new chemical processes for industry. Equally important is the ability to design the right equipment and glassware to make it possible – and safe – to carry out the experiment in the first place.

NAMES AND FORMULAS

To BE useful to chemists, chemical names should indicate the composition of a compound, as well as give structural information about the shape and the nature of the chemical bonds. But for consumers, an easy-to-remember trade name that we can associate with a known product that performs a specific job, is sufficient. Recognizing this need, chemical manufacturers developed trade names early in the development of the industry to help market their products. Many of these names do contain some chemical information and have become household words.

From the chemist's point of view, however, trade names do not give enough information, especially when referring to organic compounds that often include a huge number of complicated carbon-containing molecules. Chemists need full information about the nature of the reaction as well as the reactants and products in order to be able to derive meaning from a chemical name.

Although a number of systems of naming have been developed in the past, the majority of chemists now rely on a form of systematic nomenclature supported by the International Union of Pure and Applied Chemistry (IUPAC). In this system, the name of a compound describes various important characteristics, such as the types of cyclic compounds or functional chains the molecule contains, and what sort of bonding holds it together. The great advantage of the IUPAC system is that chemists throughout the world can understand something about the structure and chemistry of an organic compound just by knowing its name. The disadvantage is that sometimes the names are very complicated. For example, glucose (a sugar) is referred to as 1,3,4,5,6–pentahydroxy-hexanal under the IUPAC system.

Compounds can also be represented by chemical formulas, which show what elements are present in the compound and in what proportion. An example familiar even to most people outside the scientific community is the chemical formula for water, H_2O, which indicates a compound made of two atoms of hydrogen and one of oxygen.

Chemical equations offer a simple way to describe the changes that take place during chemical reactions, when a substance is converted into one or more different substances. The number of atoms of the reactants, or substances that take part in the reaction, are shown on the left-hand side of the equation. The number of atoms of the products are shown on the right-hand side. Because no atoms are either created or lost during a chemical reaction, chemical equations are balanced, so that the total number of atoms of each element on the left is equal to the total number of each on the right.

Balanced equations give a good deal of information about what compounds are made up of and how they form, but they do not indicate how quickly reactions take place. Some reactions occur spontaneously, but others require the input of energy to get them started.

The amount of energy needed to start off the breaking and re-formation of chemical bonds, and thus initiate the reaction, is known as the activation energy. The speed of a reaction can be affected by changing the temperature or pressure under which it takes place. Reactions can also be affected by a catalyst, a substance that increases the rate of a chemical reaction without actually being used up itself.

KEYWORDS

CHEMICAL EQUATION
CHEMICAL FORMULA
CHEMICAL NAME
REACTION RATE
SYSTEMATIC
   NOMENCLATURE

△ Common names can confuse. To a geologist, chalk is a rock composed mainly of small crystals; to a teacher, chalk is something to write on the blackboard with. In fact they are different, and a chemist distinguishes between them: a geologist's chalk is calcium carbonate ($CaCO_3$) and a teacher's is calcium sulfate ($CaSO_4$).

$CaCO_3$ *100 gm*
Calcium carbonate $+$ H_2SO_4 *98gm*
Sulfuric acid $\rightarrow$ $CaSO_4$ *136gm*
Calcium sulfate $+$ CO_2 *44gm*
Carbon dioxide $+$ H_2O *18gm*
Water

▽ Chemists have adopted a number of conventions for illustrating the structure of molecules.

In simple structural diagrams, bonds are represented by straight lines. Multiple bonds are represented by double or triple lines. In ball-and-stick diagrams, the sticks represent the bonds and the diagrams are drawn to give a three-dimensional perspective. In space-filling diagrams, bonds are not shown, but atoms are drawn as spheres in proportion to their relative sizes to give a clearer idea of the actual shape of the molecule. Lewis diagrams use dots to indicate the number of valence electrons of an atom. They are very useful for representing covalent bonds.

△ A chemical equation accurately describes the nature of a chemical reaction. For the example of a chemical reaction above, we can write the relative molecular masses of the reactants (on the left-hand side of the equation) and of the products (on the right). Expressing the masses of each in grams, the equation tells us that 100 grams of calcium carbonate react with 98 grams of sulfuric acid to produce 136 grams of calcium sulfate, 44 grams of carbon dioxide and 18 grams of water. This confirms that the equation balances: 198 grams of reactants yield 198 grams of products. Furthermore, we can deduce that when chalk – calcium carbonate – is dropped into acid, it effervesces as bubbles of carbon dioxide are produced.

	Oxygen	Carbon dioxide	Water	Ammonia	Methane	Benzene	Sulfur
Formula	(O_2)	(CO_2)	(H_2O)	(NH_3)	(CH_4)	(C_6H_6)	(S_8)
Space-filling							
Ball and stick							
Schematic	$O=O$	$O=C=O$	$H \quad H$ over O	$H-N$ with $H \quad H$	$H-C-H$ with H top and bottom	benzene ring	S_8 ring
Lewis	$O::O$	$O::O::O$	$H:O:H$	$H:N:H$ with H	$H:C:H$ with H	benzene Lewis	S_8 Lewis

ACIDS, BASES AND SALTS

INDIGESTION tablets use acid-base chemistry to soothe the stomach. They commonly contain bases, which react with and neutralize some of the acid present in the stomach, reducing its concentration.

Acids are compounds that contain hydrogen and dissolve in water to release hydrogen ions. Because a hydrogen ion is a hydrogen atom that has lost its electron and therefore consists of just one proton, acids are also known as proton donors. When they are dissolved in water, acids act as electrolytes and can conduct electricity.

The strongest acids are highly corrosive substances, and accordingly they must be handled with great care. Other acids, such as vinegar and citric acid, are much less corrosive. They can be used to add flavor to foods with their sour and sharp taste.

The protons in acids are readily accepted by bases. Bases are good proton acceptors because they often contain oxide (O^-) or hydroxide (OH^-) ions. Many bases cannot be dissolved in water. Those that can are known as alkalis. Indigestion tablets and cleaners for household drains are everyday bases. Industrial bases include caustic soda (sodium hydroxide, $NaOH$) and lime (calcium oxide, CaO).

When acids and bases are mixed they combine in a reaction known as neutralization to form an ionic compound, called a salt, and water. For example, hydrochloric acid (HCl) neutralizes sodium hydroxide ($NaOH$) to produce common salt (sodium chloride, $NaCl$) and water (H_2O).

KEYWORDS

ACID

ALKALI

BASE

CATALYST

CONCENTRATION

ION

pH

REACTION RATE

SALT

▷ **pH values are used to specify the strength of an acid or a base. Neutral substances, such as milk, blood and pure water, have pH values around 7. Acids have pH values of less than 7. Lemon juice (pH 2.2) and vinegar (pH 3.0) are familiar acids in the kitchen. The "sting" in stinging nettles is caused by formic acid (pH 1.5). Acid in the stomach's digestive fluids, at pH 1, is one of the strongest.**

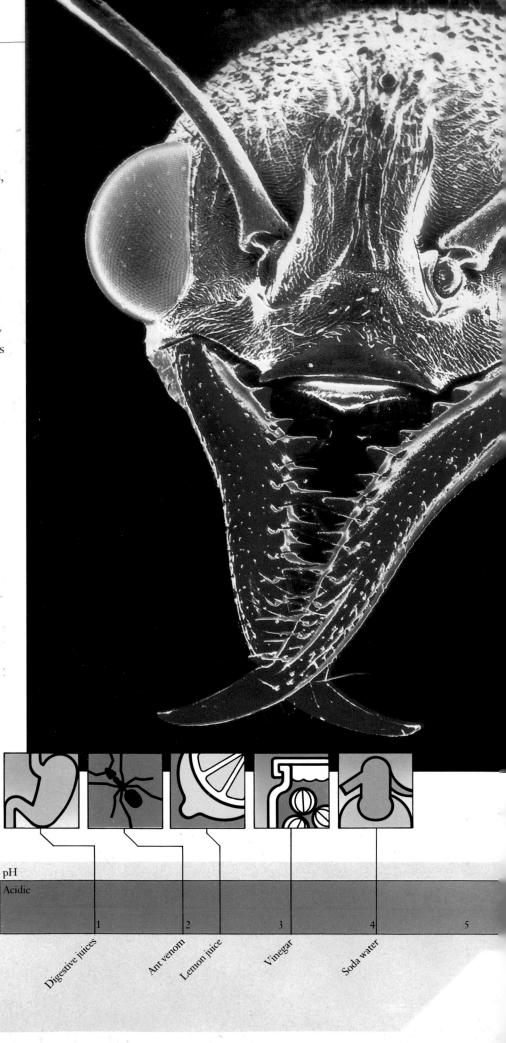

pH					
Acidic					
	1	2	3	4	5

Digestive juices   Ant venom   Lemon juice   Vinegar   Soda water

◁ The sting of an ant injects formic acid (methanoic acid, HCOOH), the simplest and most potent form of carboxylic acid. Other carboxylic acids include citric, lactic, tartaric and acetic acids. These and other acids, including benzoic and salicylic acids, are also found as salts and esters: when an acid is mixed with a water-soluble base, or alkali, the result is a salt and water.

The concentration of an acid or base varies with the amount of water present in the solution. Strength is determined by the ability to dissociate, or split up into cations and anions. Weak acids and bases only partly dissociate. Strong acids and bases, which can dissociate completely into ions, have more proton donors or acceptors.

The strength of an acid or base can be measured by the concentration of hydrogen ions in a solution, known as the pH. Various pH values can be determined with an indicator, a substance whose color changes when the pH changes. Litmus indicators distinguish between acids, which turn them red, and bases, which turn them blue. Universal indicators change color according to the pH scale.

▽ Alkaline compounds have pH values greater than 7. Toothpastes are usually slightly alkaline; the base can serve to neutralize the acids formed in the mouth by the action of bacteria on food. Cleaners, such as dishwashing detergent, often contain alkalis which act as grease removers.

△ Artists and printing houses use acid to etch fine lines in metal or glass to make plates for printing. Acids are used similarly in industry to etch small, precise patterns onto the surface of glass fiber boards and silicon chips for printed circuits and integrated circuits.

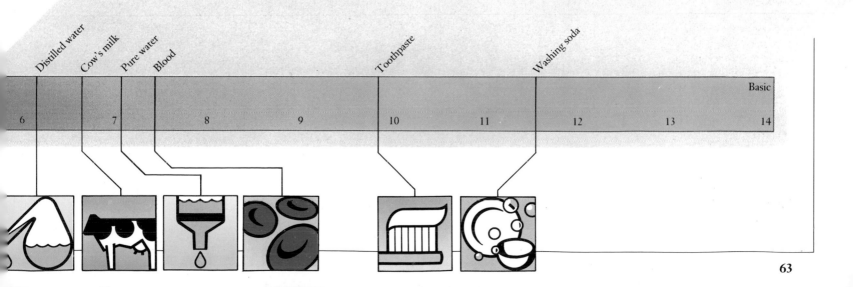

Distilled water	Cow's milk	Pure water	Blood		Toothpaste		Washing soda		Basic
6	7	8	9	10	11	12	13	14

HEAT IN AND HEAT OUT

I N ALL chemical reactions, bonds are broken, made or rearranged. During this process, energy – such as heat – is either taken in or given out. It comes from the energy changes which take place when bonds are broken and made. All reactions need some energy, known as the activation energy, to stretch and break bonds, and start the reaction off. The amount of energy needed to break a particular bond is known as its bond energy. The stronger the bond, the more difficult it is to break, and therefore the higher the bond energy is.

Reactions such as tissue respiration, neutralization and combustion – or the burning of a campfire or fireworks display – involve the formation of bonds and produce heat or release energy. These are known as exothermic reactions. Reactions such as photosynthesis (the reaction by which plants convert the energy of sunlight into food) and electrolysis (a process used to plate and purify metals) result in the breaking of bonds by taking in light or electrical energy. These are endothermic reactions. Baking a cake is another example of an endothermic reaction.

Explosions are the dramatic result when exothermic reactions produce energy faster than it can be dissipated into the surroundings, with the production of large volumes of gases. The energy given off during an exothermic reaction causes the temperature to rise, the gases to expand, and the reaction rate to increase. If the reaction accelerates quickly enough to generate a pressure wave, the explosion takes place. The controlled explosives used today depend on combustion reactions involving oxygen and a fuel.

Early explosives were very unstable because the fuel and oxidizer were combined in the same molecule. Nitroglycerin, another highly unstable substance, has been used as an explosive since the mid-18th century. It is so sensitive that it cannot be handled safely by itself, only in combination with other materials, as in dynamite. Dynamite was invented by the Swedish engineer Alfred Nobel in an attempt to find a safe way of handling liquid nitroglycerin after his brother was killed in an explosion at their factory in 1864. Dynamite requires a detonator to explode. The nitroglycerin in dynamite is absorbed in a non-volatile material: even wood pulp can be used for the purpose. Modern dynamite uses sodium nitrate to replace some of the nitroglycerin, making it even less dangerous to handle. Blasting gelatin, another modern form of explosive, contains nitrocellulose.

As long ago as 1874, the science-fiction writer Jules Verne predicted that the hydrogen and oxygen in water would one day be employed as a fuel. In fact, reactions between oxygen and hydrogen – the two components of water – are sometimes highly exothermic. A mixture of oxygen and hydrogen gas can be highly explosive, and the energy given off during the exothermic reaction which occurs when hydrogen is burned in oxygen provides the power for lift-off of spacecraft such as the US Space Shuttle. Inside the spacecraft, instead of burning hydrogen and oxygen, the two elements combine in an electric cell where they produce electricity to power the spacecraft equipment – and, equally important, water for the astronauts to drink on the voyage.

Liquid oxygen alone is used as the oxidizer in some rocket fuels. Alternatively the oxidizing agent, which enables the fuel to burn, can be another gas such as fluorine or nitrogen dioxide; a liquid with oxygen, such as hydrogen peroxide; or a solid containing oxygen such as potassium nitrate.

KEYWORDS

ACTIVATION ENERGY
BOILING POINT
BOND
COMBUSTION
ENDOTHERMIC REACTION
EXOTHERMIC REACTION
MELTING POINT
REACTION
THERMOCHEMISTRY

▷ A controlled explosion brings down a multi-storey apartment block that has been condemned. The explosive used for this operation is usually a high explosive, which is first ignited; the burning then triggers the explosion that produces a massive shock wave, causing the building to collapse. The detonation travels 1000 times faster than a flame.

◀ In the endothermic reaction that fries eggs, heat breaks the bonds that hold together proteins in the "white", allowing them to unfold and expose a greater surface area. Exothermic reactions take over as the proteins form new bonds to each other causing the egg to coagulate.

Payload

Second stage rocket

Fuel tank

Solid-fuel booster

Oxidizer tank

First stage rocket motor

△ The French Ariane V rocket and other spacecraft harness an exothermic reaction in order to lift off from the launch pad. Hydrogen, the fuel, is burned in oxygen, the oxidizing agent, producing huge amounts of heat and light. (Combustion involves the oxidation of a fuel, which means simply combining it with oxygen.) The exothermic reaction is a continuous one: it will carry on as long as there is a supply of hydrogen and oxygen. As a result of the heat generated by the reaction, the reaction product, water, takes the form of a high-pressure gas which gives the rocket the necessary thrust.

COMBUSTION AND FUEL

F UELS are compounds that contain stored chemical energy. This energy, which is released by the making and breaking of bonds, is given off in the form of heat when the fuel is broken down. The heat, in turn, can be used to do useful work, or it can be converted into other forms of energy. Food is the fuel that animals rely on to provide energy to keep them

alive. Hydrocarbons, such as oil, gas and coal, are the fuels used to provide energy to heat houses, run engines and generate electricity. Different fuels release different amounts of energy.

During respiration, organisms break down fuels such as food with oxygen to produce water (H_2O) and carbon dioxide (CO_2). The energy released as a result of this process is used to help the organism live and grow. In a similar way, fuels such as oil, gas and coal release energy when they are burned in air or oxygen to give out heat. During the combustion of hydrocarbon fuels, carbon and hydrogen react with oxygen and are oxidized to form carbon dioxide (CO_2) and water (H_2O).

The rate of combustion depends on conditions such as the concentration of oxygen. Air is only around one-fifth oxygen (the rest is mainly inert nitrogen), and fuels burn much faster in pure oxygen. Controlling the concentration of the fuel is important for controlling combustion.

Just as the energy from a match is needed to light a candle, so energy – usually in the form of heat – is needed to start off the combustion reaction. This activation energy is used to break bonds, so that new bonds can start to form. Because the combustion reaction is exothermic, it provides its own energy once the reaction gets going. As with the burning candle, the reaction stops only when the supply of fuel or oxygen runs out.

The amount of energy given off by the combustion of a fuel depends on the number of bonds to be broken and made. This is generally related to the size of the fuel molecule and the type of bonds involved. For this reason, larger hydrocarbon molecules such as hexane (C_6H_{14}), a typical constituent of gasoline, gives off more energy per molecule than fuels such as methane (natural gas, CH_4), with only a few carbon atoms. Partly oxidized fuels – including ethanol (C_2H_5OH),

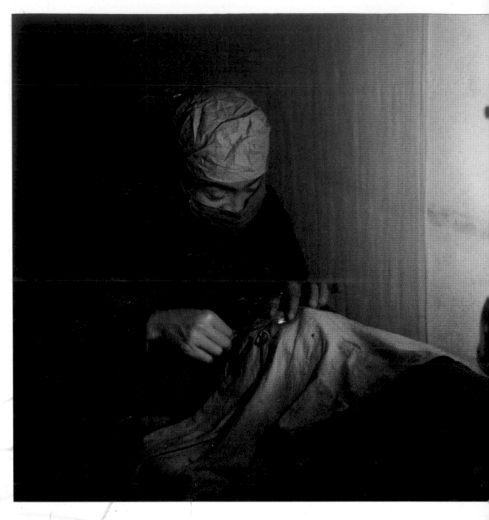

▷ **Different fuels – such as kerosene, gasoline, butane, and paraffin (wax) – give off different amounts of energy and supply varying amounts of heat. They also have different boiling and flash, or ignition, points. All of these properties are related to their individual molecular structure, especially the number of carbon atoms they contain. For example, the heavier molecules, which contain more carbon atoms, are more difficult to ignite. Gasoline (petrol), with between 5 and 10 carbon atoms in its structure, ignites at any temperature above –17°C, a useful** characteristic when it comes to starting cars. The oil used in domestic furnaces, on the other hand, which contains between 20 and 30 carbon atoms per molecule, is difficult to ignite, but once it is burning it gives off larger amounts of heat per kilogram of fuel.

Lubricating oil

Wax

◁ **Pressurized gas mantle lamps** provide a convenient portable source of illumination in places where no electricity is available. Here a surgeon uses a kerosene lamp to provide light to carry out an emergency operation in primitive conditions on a battlefield.

Butane

Gasoline

$$CH_4 + 2O_2 \rightarrow CO_2 + 2H_2O$$

Oxygen (O_2)

Kerosene   Methane (CH_4)

● Carbon

● Oxygen

○ Hydrogen

Water (H_2O)

Carbon dioxide (CO_2)

Diesel oil

▽ **The energy in all fuels** is released by the making and breaking of bonds. With hydrocarbon fuels this energy is released by burning them in oxygen to give out heat during an exothermic reaction known as combustion. In order to start off the combustion reaction, energy, usually in the form of heat, is needed to break bonds. The reaction provides its own energy once it gets going. The rate of combustion depends on the relative concentrations of oxygen, and fuel. During combustion the carbon and the hydrogen in the hydrocarbon fuel react with oxygen and are oxidized to form carbon dioxide (CO_2) and water (H_2O). Because the reaction always occurs at high temperatures, the water is in the form of vapor (steam). The carbon dioxide, however, is released as a gas into the atmosphere, where it can contribute to the "greenhouse effect".

Methane (CH_4)

Methanol (CH_3OH)

Ethanol (C_2H_5OH)

Fuel oil

Pentane (C_5H_{12})

◁ **The energy released during combustion** comes from the making of bonds to oxygen. The amount of energy given off by a particular fuel is related to the size of the fuel's molecule and the number and types of bonds that are involved. A larger hydro-carbon molecule such as pentane (C_5H_{12}), a typical ingredient in gasoline, gives off more energy than methane (CH_4), or natural gas, because it contains more bonds that can be broken to re-bond with oxygen. Fuels such as methanol (CH_3OH) and ethanol (C_2H_5OH) give off less energy because they already contain O–H bonds in their molecular structure. They are also less polluting.

alcoholic drinks – are used as alternatives to petrol in some countries, but they give off even less energy. This is because they already contain O–H bonds in their structure. Because the energy released during combustion comes from the making of bonds to oxygen, fuels that contain more oxygen give out less energy when they burn. A compromise being tested in some countries is a fuel that is a combination of an alcohol – methanol or ethanol – and conventional gasoline (petrol).

Power is not the only factor to consider when choosing a fuel. The use of alcohol-containing fuels in cars can help to reduce atmospheric pollution because they burn more completely than hydrocarbons, and give off lower amounts of carbon monoxide (CO), sulfur dioxide (SO_2) and nitrogen oxides (NO_x) when they burn. It is these compounds that combine with water in the atmosphere to form acid rain, and contribute to photochemical smog at or near ground level. The oldest fuel of all – carbon as coal or coke – is the worst culprit because its main combustion product is the greenhouse gas carbon dioxide.

CONTROLLING FIRE

Understanding the chemical reactions that occur during fires make it possible to keep a fire under control, or prevent it in the first place. Fire requires fuel, oxygen and heat. These three elements are often known as the fire triangle. Like all exothermic reactions, fires need energy to begin. Liquid fuel ignites only when it has reached the lowest temperature at which there is enough vapor to ignite – its flash point. Once burning, the fire produces heat to keep the reaction going. But if the fuel or the oxygen is removed, the fire will go out.

Fire extinguishers are designed to put out fires by attacking one or more sides of the fire triangle. There are many different types of extinguishers. The chemical and physical nature of the fuel, as well as the size of the fire, are important things to consider when deciding which to use. Water is used to put out many types of fire. It works by cooling the fire and thus breaking the heat side of the fire triangle. However, water cannot be used on electrical fires, and a stream of water directed at an oil fire serves only to disperse the oil and thus spread the fire. Some firefighting methods attempt to break two sides of the fire triangle by combining the use of water, which cools the fire, with foams, which form a heavy blanket to exclude oxygen.

Carbon dioxide (CO_2) is often used in water-based extinguishers to spray the water, but on its own, carbon dioxide is a very effective extinguisher for all confined fires. In carbon dioxide extinguishers, liquid CO_2 stored under pressure becomes a heavier-than-air gas when released and acts as an inert blanket to exclude oxygen from the fire.

Exclusion of oxygen, along with a cooling effect, is also the main action in dry chemical extinguishers, which are often used to extinguish small electrical fires. In these extinguishers, a dry powder, made up principally of sodium hydrogen carbonate (sodium bicarbonate, $NaHCO_3$), decomposes when heated to produce CO_2. As in other CO_2 extinguishers, the CO_2 gas acts to exclude oxygen, and because the reaction is driven by the heat of the fire, the CO_2 is produced exactly where it is needed. The powder itself helps to blanket and cool the fire and reduces access for air.

Halogenated hydrocarbons, such as tetrachloromethane (carbon tetrachloride, CCl_4), used in some chemical fire extinguishers have an even greater smothering effect. They are more than three times as dense as CO_2, and so provide a very effective blanket of vapor to cut off the oxygen supply. However, they must be used with care; CCl_4 decomposes to the highly toxic gas carbonyl chloride (phosgene, $COCl_2$). For this reason the use of CCl_4 extinguishers is forbidden in some places. Bromofluorocarbons have also been used in extinguishers, but are now being phased out because they are sources of bromine atoms which deplete the Earth's ozone layer.

Fire extinguishers can be useful as "fire first aid" for putting out small fires before they become dangerous. Larger fires require more than the topical application of chemicals to douse the flames. Local fire departments and other organizations often conduct short courses in fire prevention as well as in the proper and effective use of fire extinguishers.

△ A firefighter's clothing must offer protection from flames, heat and water. Modern fire-resistant clothing is often made of the synthetic fiber Aramid. This material is strong, has good insulating qualities, a high melting point and is difficult to burn. However, because Aramid is expensive, fire-resistant wool or cotton is sometimes substituted. When tightly spun and woven, these natural materials suppress burning by restricting the flow of oxygen to the fire.

◁ Bush fire threatens a house in Sydney, Australia in 1994. Large areas of vegetation become a fire hazard after a long dry season, and wind can carry the fire hundreds of miles. Such catastrophic fires are common in dry climates. Fire due to natural causes is as dangerous as fire due to an accident such as an explosion.

▷ A forest fire has to be tackled at all levels. Water from hoses and from the air is used to douse the flames, and bulldozers make fire breaks to stop the fire from spreading over the ground. Chemicals may also be sprayed from the air.

FIRE EXTINGUISHERS

All fire extinguishers work by breaking one side of the fire triangle, but they do this in different ways. CO_2 extinguishers **1** are for use on fires involving flammable liquids and electrical hazards. The vaporized gas smothers flames by displacing oxygen in the air. Dry powder extinguishers **2** are safe for use in most fires involving solids, liquids and electrical hazards. They work by smothering and cooling the fire. In soda acid extinguishers **3**, CO_2 – generated by mixing sulfuric acid with sodium carbonate – is used to propel water onto the flames. It is important to use each fire extinguisher to put out only the kind of fire for which it is designed.

Sodium carbonate solution

Sulfuric acid

A CLEANER ENVIRONMENT

WHEN plants and bacteria convert nitrogen into soluble compounds, the result is increased productivity of the soil. But when cars "fix" nitrogen, the result is air pollution. Car exhaust combines with sulfur and other emissions from factory chimneys and power plants. The result is acid rain and a dangerous cocktail of airborne chemical pollution.

Air-polluting chemicals such as oxides of carbon, sulfur and nitrogen are the byproducts of chemical industrial processes. Nitrogen oxides, along with sulfur oxides emitted from coal-fired power stations, are major causes of acid rain. Other emissions can lead to respiratory and skin problems for people who live near the plants. To cut down on emissions, plant operators can develop new processes that do not cause emissions in the first place, or they can devise ways of cleaning up the exhaust gases before they are released into the atmosphere. Most operators of factories choose the clean-up option.

For factory chimneys, flue gas desulfurization is the most common method for neutralizing acidic gases such as sulfur dioxide (SO_2). Typically this process uses scrubbers, compartments inserted inside the chimney in which an alkali such as limestone reacts with the sulfur dioxide and removes it in the form of gypsum, which can be used in other ways.

Cars, trucks and buses are by far a greater source of pollutants. Vehicle exhausts are responsible for most of the carbon monoxide (CO) released into the air. They also release nitrogen oxides, as well as unburned hydrocarbons. When nitrogen oxides react with unburned hydrocarbons, oxygen and water vapor in the presence of sunlight, ozone is formed near the ground. Ozone in the upper atmosphere protects the Earth from harmful ultraviolet radiation, but at ground level it damages many biological molecules, and results in eye-irritating photochemical smogs.

Vehicle exhaust pollution is caused by the incomplete combustion of fuel because there is insufficient oxygen. Automotive engineers are currently experimenting with "lean burn" engines which use a higher air-to-fuel ratio than normal engines in order to burn fuel more completely.

However, the most efficient way to cut down on vehicle exhaust emissions is to fit a three-way catalytic converter between the engine and exhaust pipe. It promotes chemical reactions that change carbon monoxide, unburned hydrocarbons and nitrogen oxides into carbon dioxide (CO_2), water and nitrogen, by promoting the oxidation of carbon monoxide and hydrocarbons, and the reduction of nitrogen oxides. In the converter, a catalyst such as powdered platinum and rhodium is spread over a ceramic support. In some catalytic converters, the converter consists of a packed bed of small porous ceramic spheres whose outer shell is impregnated with the catalyst metal. The catalysts can be used only with lead-free fuel.

▶ Flue scrubbers are used to remove sulfur from factory emissions. The reaction requires no energy input, but uses lime (calcium oxide, CaO), an alkali made by heating limestone ($CaCO_3$) to produce $CaO+CO_2$. In the scrubber, acidic SO_2 gas is reacted with oxygen in the air to form solid calcium sulfate ($CaSO_4$), or gypsum. The desulfurized gases are emitted into the atmosphere. The gypsum is recovered and used in the manufacture of plaster.

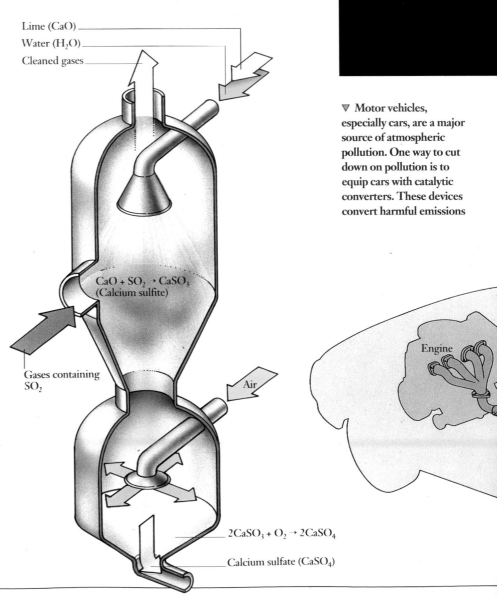

Lime (CaO)

Water (H_2O)

Cleaned gases

$CaO + SO_2 \rightarrow CaSO_3$
(Calcium sulfite)

Gases containing SO_2

Air

Engine

$2CaSO_3 + O_2 \rightarrow 2CaSO_4$

Calcium sulfate ($CaSO_4$)

▽ Motor vehicles, especially cars, are a major source of atmospheric pollution. One way to cut down on pollution is to equip cars with catalytic converters. These devices convert harmful emissions

◁ Factory emissions can lead to breathing and skin problems for people who live near the plants. They also lead to acid rain, which can cause extensive damage to plants, animals and buildings far from the factory site.

▽ Photochemical smog (a combination of fog and toxic smoke, soot and ozone) and carbon monoxide (from the exhaust fumes of cars and trucks) dim the morning sunlight in Mexico City, one of the most polluted in the world.

such as nitrogen oxides (NO_x), carbon monoxide (CO) and unburned hydrocarbons (C_xH_x) into nitrogen (N_2), carbon dioxide (CO_2) and water (H_2O), all of which are naturally present in air.

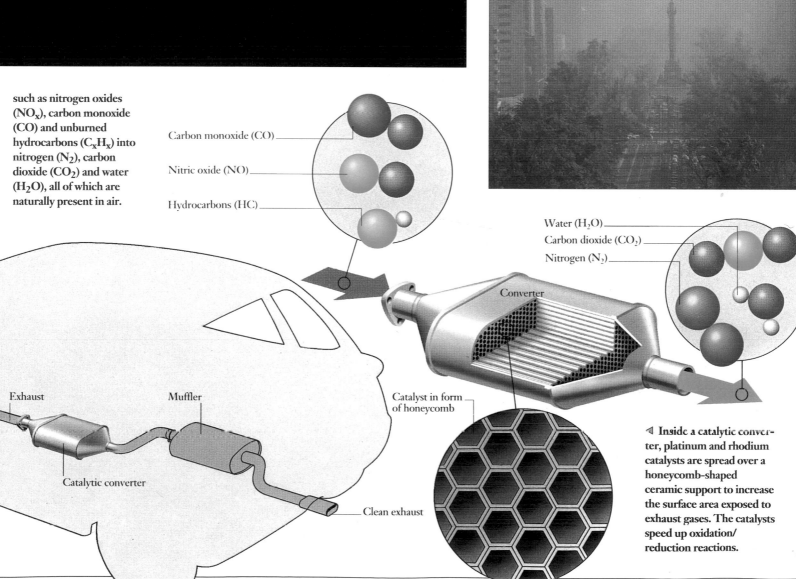

Carbon monoxide (CO)

Nitric oxide (NO)

Hydrocarbons (HC)

Water (H_2O)
Carbon dioxide (CO_2)
Nitrogen (N_2)

Converter

Exhaust

Muffler

Catalytic converter

Clean exhaust

Catalyst in form of honeycomb

◁ Inside a catalytic converter, platinum and rhodium catalysts are spread over a honeycomb-shaped ceramic support to increase the surface area exposed to exhaust gases. The catalysts speed up oxidation/reduction reactions.

ELECTRICITY AND CHEMISTRY

T HE addition and removal of electrons is one of the key processes in chemical bonding – and it is the basis of electrochemistry, in which electrical potentials are used to drive chemical reactions.

To industrial chemists, electrochemistry offers many advantages. For a start, it can be used to produce many products more cheaply, and with less harm to the environment. This is because the energy needed to drive electrochemical reactions comes largely from the electrical potential of the reactants. As a consequence, reactions can be run successfully at lower temperatures, and energy costs are lower. In addition, because only the direct transfer of electrons is involved, many oxidation/reduction processes can be carried out without using harsh chemicals. Instead, electrons are supplied directly to the chemical reaction via electrodes dipped into the reaction mixture.

Electrochemical reactions are carried out in electrochemical cells. These contain an electrolyte, a power supply, a positive electrode (anode) which accepts electrons, and a negative electrode (cathode) which releases electrons. The electrolyte is an ionic compound, such as an acid, alkali or salt, which dissolves in water to release ions and allows ions to flow through it. Electrolytes conduct electricity when they are melted or dissolved, but not necessarily when they are solid. An electrochemical cell works in the opposite way to a battery. In a battery, electricity is produced by a chemical reaction: the movement of ions through an electrolyte from one type of metal electrode to another. By varying the different types of metals used it is possible to produce different voltages. In an electrochemical cell, electricity is used to drive a chemical reaction.

Electrochemistry is already used in a wide range of industrial processes. Many important chemicals such as sodium hydroxide and chlorine are manufactured in electrochemical cells. Electrolysis, the process which occurs when an electrolyte conducts electricity, is also used to extract reactive metals, refine metals and anodize aluminum to give a tougher surface. In electroplating, electrolysis is used to coat metals. Electrolysis is also a useful method of purifying metals such as copper. Here, impure copper is used as the anode, and a sheet of pure copper is used as the cathode. The electrolyte is an acidified copper (II)

KEYWORDS

ANODE

CATHODE

ELECTROLYSIS

ELECTROLYTE

ELECTROPLATING

HALL-HÉROULT PROCESS

ION

OXIDATION

REDOX REACTION

▽ Computers and other electronic equipment contain enough gold and other valuable metals in their circuitry to make recovery and recycling worthwhile. The recovery process is similar to electroplating. However, in electroplating, the object to be plated is used as the cathode, while the anode is made up of the metal which is used as plating. In deplating, the object to be stripped is used as the anode, and the recovered metal is deposited on a cathode, usually of the same metal.

sulfate solution ($CuSO_4$). During electrolysis the anode dissolves and the copper ions are deposited as pure copper on the cathode.

Chemists are now beginning to exploit electrochemistry in new ways. Because chemical changes in an electrochemical cell involve the movement of ions which can be easily measured by measuring electrical current, electrochemical cells have proved to be a useful tool to monitor reactions. In medicine, electrochemical cells can be used as biological monitoring devices, or biosensors, by adapting them to monitor a specific biological reaction. For example, a tiny glucose sensor for use by diabetics has been developed. It monitors glucose levels by measuring the change in potential difference which occurs when glucose reacts with a specific enzyme.

Electrochemistry is also being explored as an alternative means of destroying organic waste materials. Oxidation is necessary to break down the organic waste molecules into harmless water and carbon dioxide. The oxidation is often carried out by burning the waste in an incinerator. Chemists are investigating the use of electrochemically-generated, highly oxidizing metal ions to carry out the oxidation at much lower temperatures, without the risk of releasing harmful gases as burning does.

△ Hundreds of mercury cells in a chlorine plant, in which electrolysis of brine produces three industrial products: sodium hydroxide (NaOH), chlorine and hydrogen gas. The method involves graphite anodes and a pool of liquid mercury as a cathode; the electrolyte is brine, made up of 25 percent salt (NaCl) dissolved in water. Chlorine gas is given off at the graphite anodes. The sodium atoms discharge at the negative mercury cathode, where they dissolve in the mercury to form an amalgam. This amalgam flows out into another cell, where it is mixed with water over activated carbon to remove the sodium. The mercury flows back into the original cell, while the sodium in the second cell reacts with the water to produce sodium hydroxide solution and hydrogen gas.

△ Aluminum and its alloys are used for many purposes, ranging from the mundane to the exotic. Aluminum is a familiar component of drinks cans, food packaging and pots and pans. It can also be used in sculpture.

▽ When an electric current is passed between the electrodes, positive aluminum ions collect at the negatively-charged cathode, where they form molten aluminum which can be drained off. The process consumes large amounts of electricity, so it is economical only where a reliable supply of cheap electricity is available. Feldspar, clay and many other minerals are also sources of aluminum.

▽ Aluminum – the most abundant element in the Earth's crust – is commonly extracted from its main ore, bauxite, using the electrolysis-based Hall-Héroult process. In this process, ore is dissolved (at a temperature of 1000°C) in molten cryolite, a naturally occurring aluminum fluoride salt which acts as an electrolyte. Carbon is used for both the anodes and cathodes. Electrolysis is carried out in a special carbon-lined tank which acts as the cathode.

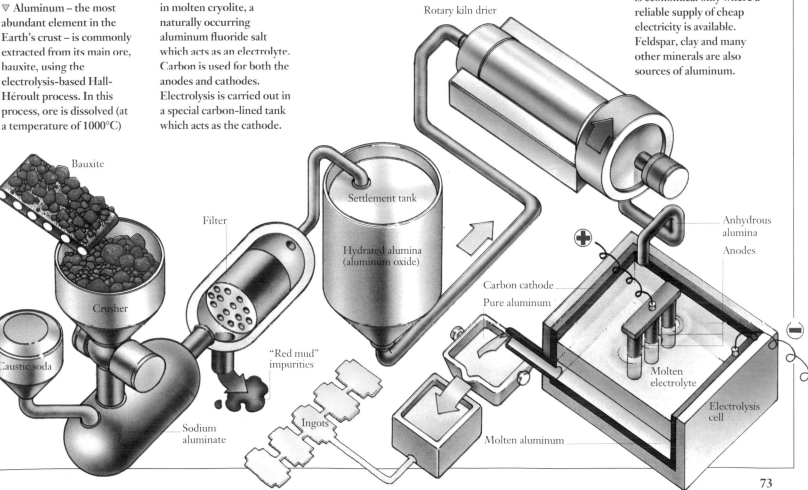

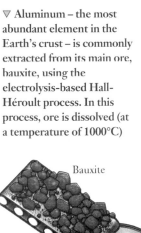

Rotary kiln drier

Bauxite

Filter

Settlement tank

Hydrated alumina (aluminum oxide)

Anhydrous alumina

Anodes

Carbon cathode

Pure aluminum

Molten electrolyte

Crusher

Caustic soda

"Red mud" impurities

Sodium aluminate

Ingots

Molten aluminum

Electrolysis cell

THE CHEMICAL INDUSTRY

THE chemical industry works to transform common raw materials such as oil, gas, coal, minerals, air and water cheaply and efficiently into chemicals that can be used in the manufacture of other things. At a chemical plant, the reactants – or feedstock – are combined under appropriate conditions to produce the desired product. But simply using chemical reactions to turn one substance into another is not enough. It is important to find ways to make reactions occur more efficiently, and to increase the reaction rate which is measured in terms of the change in concentration of a reactant or product over time. This can sometimes be achieved by carrying out reactions at higher temperatures or pressures, increasing the concentration of the reactants, or using catalysts which make it easier for the reactants to react with each other at lower temperatures. The aim is to obtain the optimum yield, calculated as a percentage of amount of product/amount of feedstock used to produce it.

However, the optimum yield is not necessarily the maximum yield that can be achieved. When designing a plant, chemical engineers must consider the costs of maintaining the high pressures and temperatures needed to maximize the reaction rate for some products, and the fact that fast exothermic reactions can be very difficult to control. Plant operators and chemical engineers must weigh up safety and energy costs when choosing the reaction conditions.

Chemical engineers must also determine which type of processing to use. In a batch process, raw materials are put into a vessel and allowed to react. When the reaction is complete, the product is removed and new feedstock is added to make the next batch. In a continuous process, feedstock is constantly fed into the plant where it passes through and reacts to give a continuous flow of that product.

Batch processes are useful for slow reactions which produce relatively small amounts of product. Pharmaceuticals and cosmetics are typically manufactured in this way. They are also useful for manufacturing products where there is an explosion risk, or for fermentation processes where there is a risk of contamination.

Continuous processes are an efficient means of high-tonnage production. For example, the important industrial chemicals ammonia and chlorine are

manufactured in this way. However, continuous processes require tailor-made plants, and these are expensive to set up.

Waste is an inevitable product of all chemical plants. In the past chemical manufacturers gave little thought to waste disposal, and as a result chemical plants were notorious for the large amounts of pollution and contamination they created. Now attitudes are changing, and waste management is an important consideration in all plant design.

In the past, operators tended to favor the "dilution solution" – dumping waste into the atmosphere or into rivers, lakes or oceans with the hope that it would become sufficiently diluted so as not to be harmful. Now waste is sometimes contained in purpose-built ponds or heaps. However, the waste can contaminate the land or leak into groundwater or rivers. Chemical manufacturers are developing chemical and mechanical waste treatment methods to meet the tougher legal requirements for waste disposal.

▲ The Haber process is a continuous reaction used to make ammonia (NH_3) on an industrial scale. A typical plant produces 1200 tonnes of ammonia each day.

▷ In the Haber process, nitrogen (N_2) is combined with hydrogen (H_2) to make ammonia (NH_3). Although the original process used hydrogen derived from water (H_2O), the hydrogen is now obtained by reacting steam with natural gas, methane (CH_4). In the process, nitrogen derived from air is mixed with hydrogen, and the heated gases pressurized and passed over a catalyst.

◁ The modern process of brewing is an example of a batch process which has been carried out in a similar way for thousands of years. In brewing, the use of batch processing helps to cut down on the risk of contamination by unwanted microbes. Generally, batch processes are suitable for producing relatively small amounts of product using slow reactions.

▽ The most common use of ammonia is for agricultural fertilizers, in which it provides a source of the nitrogen needed by plants for healthy growth. The nitrogen may be applied in the form of NH_3 by injecting ammonia directly into the soil or by converting it into nitrate fertilizers, which can then be spread by tractors, as shown here.

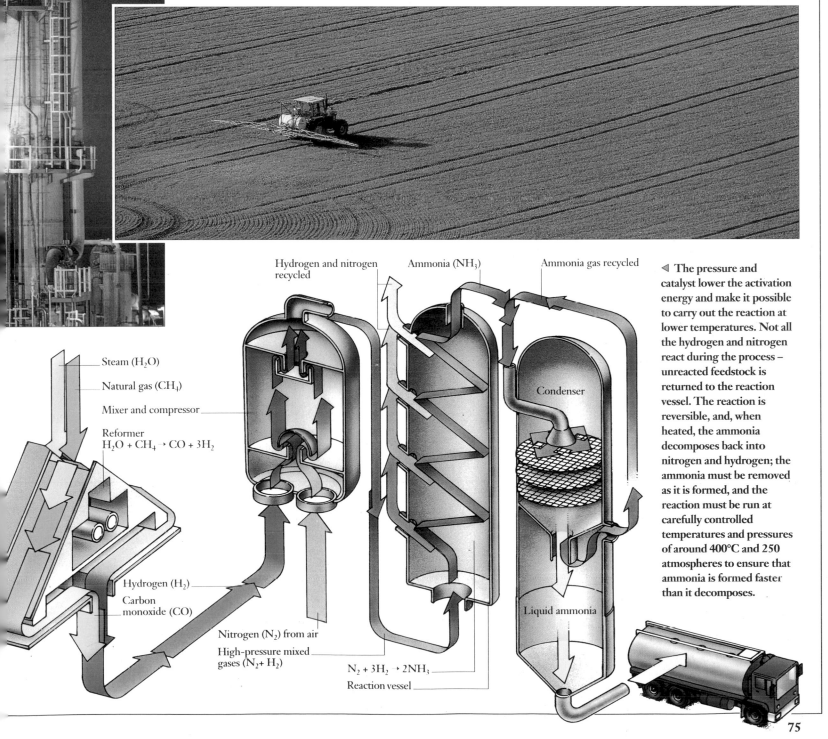

Steam (H_2O)

Natural gas (CH_4)

Mixer and compressor

Reformer
$H_2O + CH_4 \rightarrow CO + 3H_2$

Hydrogen (H_2)

Carbon monoxide (CO)

Nitrogen (N_2) from air

High-pressure mixed gases ($N_2 + H_2$)

Hydrogen and nitrogen recycled

Ammonia (NH_3)

Ammonia gas recycled

Condenser

Liquid ammonia

$N_2 + 3H_2 \rightarrow 2NH_3$

Reaction vessel

◁ The pressure and catalyst lower the activation energy and make it possible to carry out the reaction at lower temperatures. Not all the hydrogen and nitrogen react during the process – unreacted feedstock is returned to the reaction vessel. The reaction is reversible, and, when heated, the ammonia decomposes back into nitrogen and hydrogen; the ammonia must be removed as it is formed, and the reaction must be run at carefully controlled temperatures and pressures of around 400°C and 250 atmospheres to ensure that ammonia is formed faster than it decomposes.

MAKING ACIDS AND BASES

SOME economists claim that the amount of sulfuric acid produced by a nation is a good indicator of its economic health. Whether true or not, there is no doubt that the manufacture of acids and bases for use in a wide range of industrial processes is one of the most important activities of the chemical industry.

Sulfuric acid (H_2SO_4) is probably the leading product of the chemical industry, and most of the sulfur produced worldwide is used for its manufacture.

In the United States, for example, almost twice as much sulfuric acid is produced as any other chemical.

Sulfuric acid is one of the strongest acids known and has a wide range of applications in almost all manufacturing processes. It is commonly used in the manufacture of dyes, paints, paper pulp, explosives, car batteries and fertilizers. It is also used to make some detergents and in petroleum and metal refining. It is an excellent dehydrating agent, and dissolves many metals to form a wide variety of industrial compounds.

Sulfuric acid is manufactured by the contact process. Here sulfur is first burned in dry air to form sulfur dioxide (SO_2) gas. The hot SO_2 is reacted with more oxygen at a temperature of around 450°C and in the presence of a vanadium oxide (V_2O_5) catalyst to form sulfur trioxide (SO_3). Sulfuric acid can be formed by reacting the SO_3 with water, but this reaction is very violent and forms a mist of sulfuric acid droplets which is difficult to absorb. Instead, SO_3 is dissolved in concentrated (98 percent solution) sulfuric acid, where it reacts less violently with the small amount of water present and does not form a mist. The resulting product is oleum ($H_2S_2O_7$), which is used in some industrial processes. To make sulfuric acid, the oleum is diluted with water to form concentrated sulfuric acid. Alternatively, as the SO_3 is added to the concentrated sulfuric acid, some acid is continuously removed, and water is added to the remainder to keep its concentration at around 98 percent.

▷ **To extract sulfur, super-heated water is pumped down the outer pipe under pressure, and compressed air is forced down the inner pipe. The hot water melts the sulfur, which accumulates at one end, then rises up the middle pipe.**

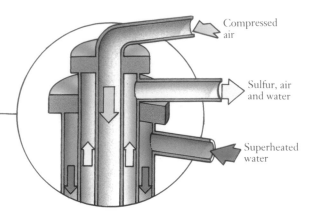

Rock

Calcite

Air

Hot water

Melted sulfur

△ Sulfur can be produced commercially by extraction from crude oil and raw natural gas. However, in some parts of the world, sulfur occurs naturally underground in association with layers of calcite ($CaCO_3$). This form of sulfur is extracted using the Frasch process.

▽ Sulfuric acid has many uses in industry. Steel is dipped in sulfuric acid to clean or "pickle" it in preparation for galvanizing (protecting with zinc coating, shown here). Sulfuric acid is an oxidizing and dehydrating agent, and reacts on heating with metals, sulfur and carbon.

Compressed air

Sulfur, air and water

Superheated water

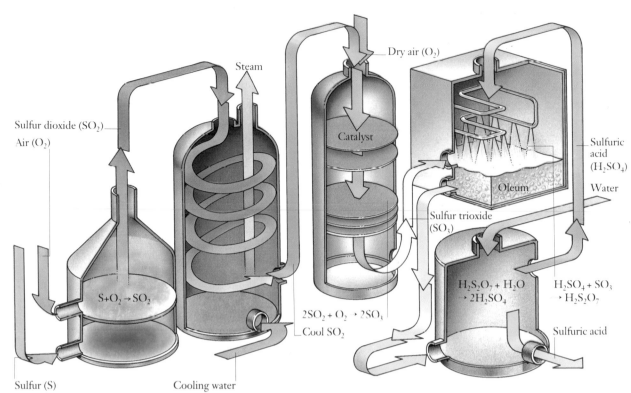

◁ The contact process is used to manufacture sulfuric acid (H_2SO_4). First sulfur dioxide (SO_2) gas is formed by burning dry sulfur. At a temperature around 450°C, the hot gas oxidizes (reacts with oxygen) to form sulfur trioxide (SO_3), using a catalyst of vanadium oxide or platinum. The sulfur trioxide is then dissolved in concentrated sulfuric acid to form a product known as oleum ($H_2S_2O_7$) – a thick, oily liquid. This may be diluted with water to give a 98 percent solution of sulfuric acid, which is thinner and less oily.

Steam

Dry air (O_2)

Sulfur dioxide (SO_2)

Air (O_2)

Catalyst

Sulfuric acid (H_2SO_4)

Oleum

Water

Sulfur trioxide (SO_3)

$S+O_2 \rightarrow SO_2$

$2SO_2 + O_2 \rightarrow 2SO_3$

Cool SO_2

$H_2S_2O_7 + H_2O \rightarrow 2H_2SO_4$

$H_2SO_4 + SO_3 \rightarrow H_2S_2O_7$

Sulfuric acid

Sulfur (S)

Cooling water

Among the bases, caustic soda (sodium hydroxide, NaOH) is one of the most important. It is widely used in processes including the manufacture of soap, paper, detergents and other chemicals as well as for the production of rayon and acetate fibers.

Caustic soda is produced in chloralkali plants by the electrolysis of brine, a concentrated solution of sodium chloride (NaCl), or common salt, in water that contains about 25 percent by mass NaCl. In order to be economic, the plants must be located near a good source of salt and a cheap supply of electricity, because the power consumed by the process is very high – equivalent to the output of a large modern power station, or the requirements of a large town.

In the process, electricity is passed through the brine in an electrolytic cell, which includes a porous diaphragm to separate the anode and cathode compartments and prevent reaction of their products. Sodium collects at the cathode where it reacts with water to form sodium ions and release hydrogen gas, and hydroxide ions. As the reaction proceeds a mixed solution of NaCl and NaOH is formed. This solution is concentrated in large evaporators, where the NaCl crystallizes out and is removed by filtration.

During the process chlorine gas evolves at the anode, where it is cooled to condense out most of the water and "scrubbed" with sulfuric acid to make a saleable product in its own right. It is widely used for the production of the plastic polyvinyl chloride (PVC).

Hydrogen is another product of the reaction. The hydrogen gas can be sold as a commercial product, either for the hydrogenation of fats, or it can be used with the chlorine to manufacture hydrochloric acid.

MAKING USEFUL CHEMICALS

COMMON salt (sodium chloride, NaCl) is an essential starting point for the production of many industrial chemicals. Not only is it widely used in the production of caustic soda (sodium hydroxide, NaOH); common salt also forms the basis of another important industrial alkali, sodium carbonate ($NaCO_3$), also known as soda ash or washing soda. Sodium carbonate has a wide variety of applications. It is used as a source of alkalinity in boiler water to help prevent corrosion, as a water softener, and as an ingredient in soaps, detergents and other cleansing agents, and in photographic developers. Sodium carbonate is also used in steel processing, enameling, textiles, dyes, food and drink, and in the processing of oils, fats, waxes and sugars. It is also important in the commercial manufacture of glass.

In some parts of the world sodium carbonate is obtained by mining and purifying trona, an evaporite mineral deposited as the result of the evaporation of ancient seas, which contains sodium carbonate and sodium hydrogen carbonate (bicarbonate) and a few impurities. Where trona deposits are not available, synthetic sodium carbonate is manufactured using the Solvay, or ammonia-soda, process.

The overall reaction used in the Solvay process involves the combination of sodium chloride and calcium carbonate to produce sodium carbonate and calcium chloride. The process includes a number of intermediate steps and requires the input of ammonia, generally manufactured using the Haber process, to prevent the favored reverse reaction from taking place.

During the Solvay process a number of useful intermediate products are also produced. The only intermediate product that could be considered waste is calcium chloride ($CaCl_2$), but even this may be used as an energy storage fluid for solar heating systems.

The more widely used intermediate products include ammonium chloride, which is often recycled back into the reactor to provide additional ammonia, and sodium hydrogencarbonate ($NaHCO_3$), also known as baking soda or sodium bicarbonate. $NaHCO_3$ is widely used as a raising agent in baking, an antacid agent in indigestion medicines and as the soda in acid-soda fire extinguishers. It can also be heated to produce a light ash which, when treated with water, produces more crystals of sodium carbonate.

Glass is made by heating sand (SiO_2), limestone ($CaCO_3$), sodium carbonate, cullet (broken waste glass) and small amounts of metal oxides. The metal oxides lower the melting point and also alter other characteristics of the glass, including color. Cobalt oxide is added to produce a blue glass; manganese gives a purple color; copper produces either red or blue-green glass; and chromium gives a green color.

The addition of oxides can also be used to give the glass other characteristics. Lead oxide gives the glass a high refractive index and makes it sparkle because of the greater internal reflection of light, ideal for use in cut glass. Borosilicate glass (Pyrex), a heat-resistant glass which is also resistant to most chemical attack, is made by adding 10-15 percent boron oxide (B_2O_3). Photochromatic sunglasses, which turn darker in the light and lose their color in the dark, are made by incorporating silver chloride (AgCl), the active ingredient in photographic film, to glass.

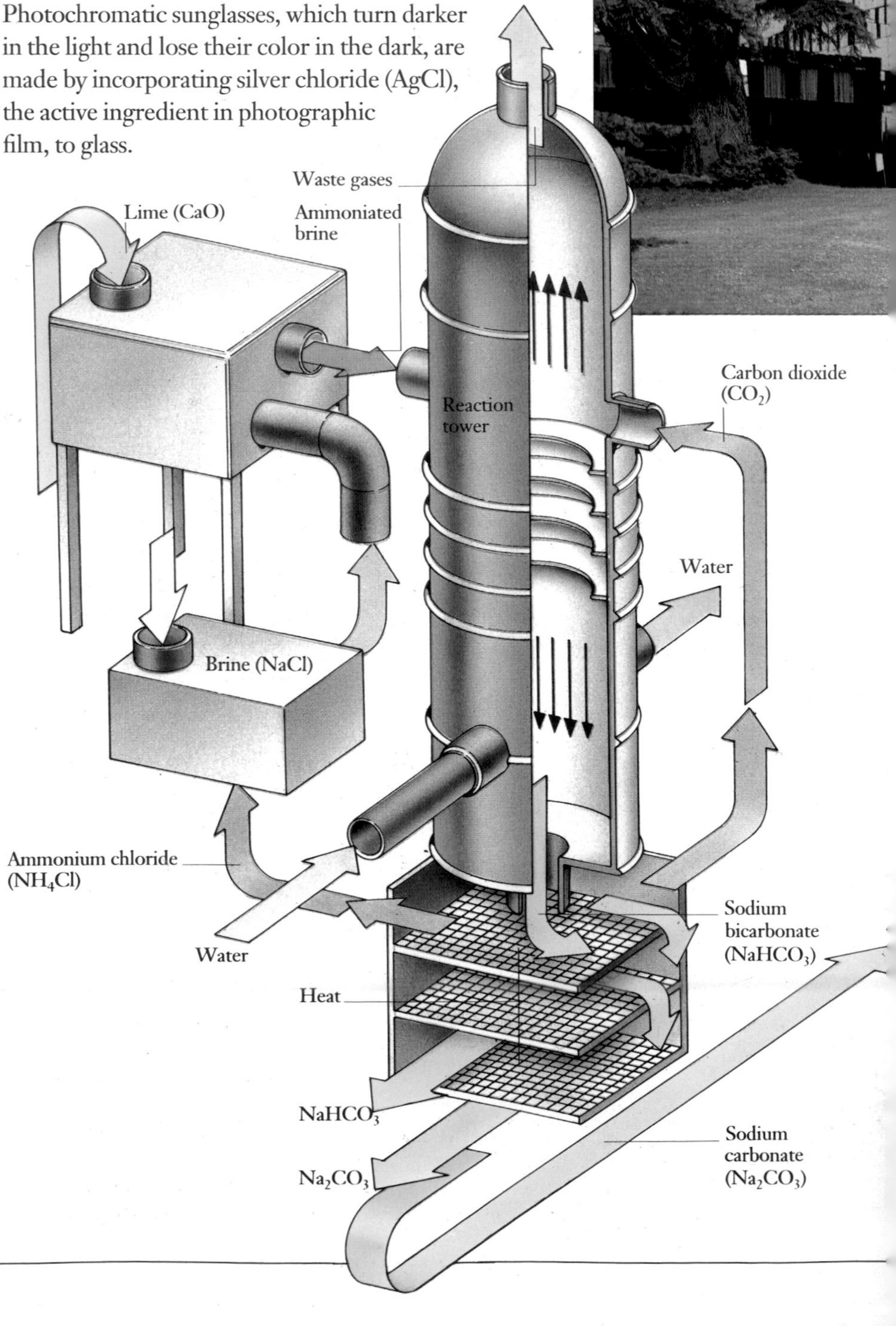

Waste gases

Lime (CaO)

Ammoniated brine

Carbon dioxide (CO_2)

Reaction tower

Water

Brine (NaCl)

Ammonium chloride (NH_4Cl)

Water

Heat

Sodium bicarbonate ($NaHCO_3$)

$NaHCO_3$

Na_2CO_3

Sodium carbonate (Na_2CO_3)

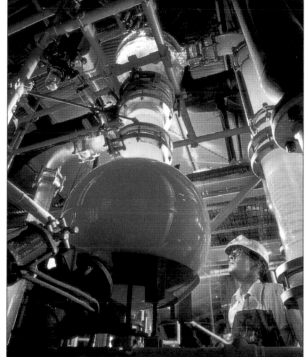

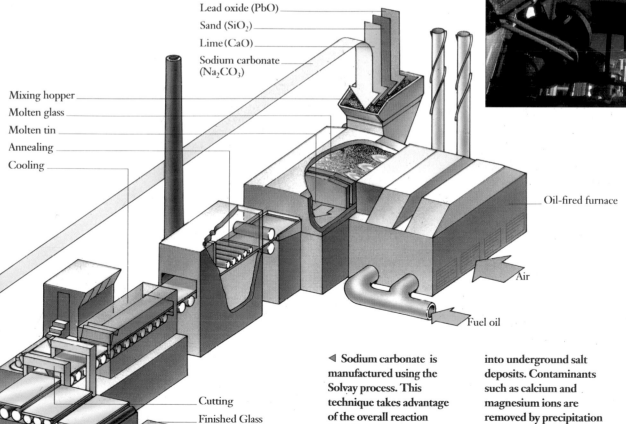

◄ Glass has taken its place as a modern building material. It cannot be used as a structural material, but is increasingly used as a cladding. Glass cladding lets in a great deal of natural light, but can be coated to reflect the Sun's rays and thus prevent overheating.

▽ Glass can be made resistant to heat and attack from almost all chemicals except hydrofluoric acid (HF) by the addition of small amounts of boron oxide (B_2O_3). Because it is so resistant, vessels made of borosilicate glass are often used in the batch production of chemicals.

Lead oxide (PbO)
Sand (SiO_2)
Lime (CaO)
Sodium carbonate (Na_2CO_3)

Mixing hopper
Molten glass
Molten tin
Annealing
Cooling

Oil-fired furnace

Air

Fuel oil

Cutting
Finished Glass

◄ Sodium carbonate is manufactured using the Solvay process. This technique takes advantage of the overall reaction $2NaCl=CaCO_3 \rightarrow Na_2CO_3 +CaCl_2$. A brine solution is formed by injecting water into underground salt deposits. Contaminants such as calcium and magnesium ions are removed by precipitation with sodium hydroxide (NaOH) or sodium carbonate (Na_2CO_3), and

the brine is saturated with ammonia gas and carbon dioxide obtained by roasting limestone. The decontaminated brine is then passed down tall towers. Plates inside the towers break up the gas stream and ensure efficient mixing with the brine. The ammonia and carbon dioxide react to form ammonia hydrogen carbonate (NH_4HCO_3), which reacts with NaCl to form sodium hydrogen carbonate (sodium bicarbonate, $NaHCO_3$). This now precipitates out, and is converted to sodium carbonate (Na_2CO_3) by heating to release carbon dioxide and water.

SOAPS AND DETERGENTS

CLEANING products were among the earliest chemicals produced commercially. Soaps were originally manufactured by boiling animal fat and lye, the alkali leached out of wood ash. Today the lye is replaced by alkali caustic soda (sodium hydroxide, NaOH), and vegetable oils replace the animal fat, but the basic chemistry remains the same.

For many purposes, detergents may be used in place of soap. Hard water contains calcium and magnesium compounds, which form a scum with soap; detergents do not form scum. Scum is not only messy, it also wastes soap, because soap does not lather until it has reacted with all the dissolved substances in the water.

Soaps are produced by a chemical reaction called saponification. An ester – a compound formed by the reaction of an acid with an alcohol – reacts with caustic soda to produce soap and the original alcohol. Even now, as in the past, most soaps continue to be made from natural oils and fats, whereas detergents are produced from hydrocarbons extracted from petroleum.

Soap is a mixture of the salts of long-chain carboxylic acids -- organic acids that have a hydrophilic, or water-soluble, polar carboxyl group (–COOH) at one end (the head), and a

> Soaps and detergents work to clean materials in three ways. First, they improve the wettability of water by lowering the surface tension at the interface between the water and the material to be cleaned. Second, they make it possible for grease molecules to dissolve in water. Third, they keep the grease in suspension so that it can be rinsed away.

KEYWORDS

DETERGENT

ENZYME

ESTER

FOAM STABILIZER

HYDROPHILIC

HYDROPHOBIC

LONG-CHAIN
  HYDROCARBON

SAPONIFICATION

SOAP

SURFACTANT

WETTABILITY

▷ The molecular structures of soaps and detergents are similar. Both have a water-loving (hydrophilic) ionic group at one end (usually called the head), with a long grease- or fat-loving (hydrophobic) "tail" made up of various organic groups.

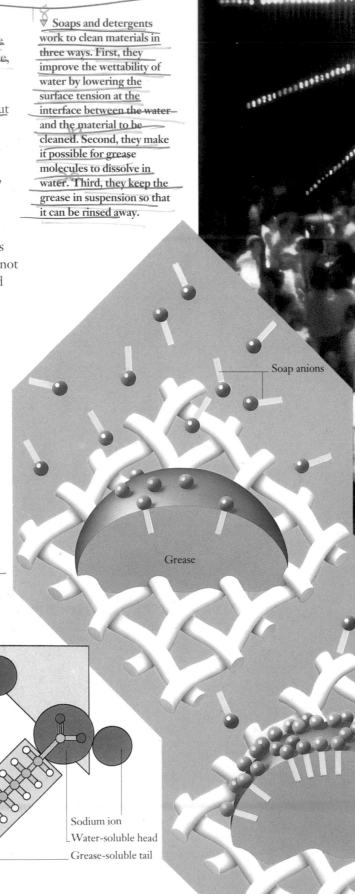

Soap anions

Grease

Detergent

Sodium ion

Water-soluble head

Grease-soluble tail

Soap

◁ Under ultraviolet lights at a disco, fabrics washed in detergents with optical brighteners glow in the dark. Optical brighteners are fluorescent substances whose molecules absorb light at one wavelength and emit it at another.

△ A seabird covered in oil after an oil spill at sea can have its feathers cleaned by a detergent. The detergent breaks the oil up into small droplets which can then be washed away by water. Detergents can be used in large volumes to break up oil slicks on the sea surface.

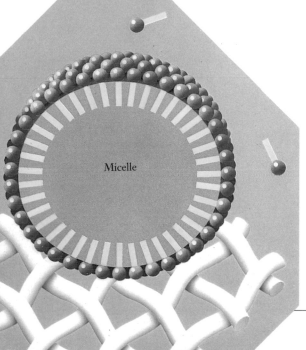

Micelle

◁ When a soap or detergent goes to work on a piece of greasy cloth, its molecules plug their hydrophobic tails into the droplets of grease attached to the fibers. Eventually the soap molecules surround the droplets to form spherical micelles, which float off the cloth and can be washed away. The detergent also acts as an emulsifier and carries the grease away in solution.

(–COOH) at one end (the head), and a hydrophobic, or water-insoluble, nonpolar grease-loving group at the other (the tail). In chemical terms, soaps are defined as the sodium salts of long-chain fatty acids, and have the general formula $RCOO^-Na^+$, where R is a long hydrophobic hydrocarbon chain.

Like soaps, detergent molecules have a fairly long nonpolar tail and a polar head. But in detergents, also known as synthetic surfactants, the nonpolar tails are produced by a series of chemical reactions on hydrocarbons. For a typical detergent, four molecules of propene (propylene, $CH_3–CH=CH_2$) are joined to a benzene ring attached at the double bond, and reacted with a sulfonic acid to form a branched-chain alkylbenzene sulfonate. By altering the chemical composition, the properties of a detergent can be tailor-made for different cleaning jobs. Modern laundry detergents include foam stabilizers; water softeners to counteract the activity of calcium and magnesium ions in hard water; and organic binders, such as sodium carboxymethyl cellulose, to ensure dirt stays in suspension. These work by increasing the negative charge in fabrics which then repel the negatively charged dirt particles. Some laundry detergents contain "optical brighteners". These substances are molecules that fluoresce – absorb light at one wavelength and emit it at another – to give a blue or ultraviolet light. Optical brighteners overcome any yellow tinge in the fabric, restore the mixture of colors to those that a white fabric would normally reflect, and – in advertising terminology – make white fabrics look "brighter than bright". Biological laundry detergents also contain enzymes – biological catalysts that break down and "digest" substances such as proteins, blood and sweat.

For use as a shampoo, a detergent must be designed to act rather differently. Shampoos must remove oil and dirt while at the same time leaving a thin coating of the natural oil sebum; if too much sebum is washed away, the hair will dry out. Shampoos for greasy hair are designed to remove more of the sebum, whereas for dry hair they are designed to remove less. Shampoo for dry hair often contains fatty material to supplement the natural sebum in the hair.

The acidity of the detergent is also important. Shampoos must be designed to be neutral (with a pH of around 7), because more alkaline compounds, with a higher pH, make the individual strands of hair break easily, and ruffle the outer surface of the hair, which can make it look dull and coarse.

3

ORGANIC
Chemistry

MILLIONS OF YEARS AGO, carbon compounds in the
atmosphere formed an insulating blanket around the
Earth which trapped the Sun's heat. This gradually
made the Earth warm enough for life to evolve. Carbon remains
central to life on Earth, although it makes up less than 1 percent of
the Earth: the molecules that make up all living things are based on
carbon compounds. Biochemistry, the chemistry of living organisms,
is primarily concerned with carbon compounds.

Carbon circulates through plants, animals, the soil and the
atmosphere in a process known as the carbon cycle. Carbon is also
returned to the environment when biomass fuels such as wood, or
fossil fuels such as oil, gas and coal – made up of carbon-containing
organic material that originated millions of years ago – are burned to
release energy. Carbon-based fuels account for 75 percent of the
energy that is used on the planet today.

Carbon is versatile because each carbon atom forms four covalent
bonds. Organic molecules (so called because chemists once thought
that these compounds could only be found in living organisms) are
made up of carbon atoms bonded together by single, double or triple
covalent bonds. The range of compounds formed by these bonded
carbon atoms are the basis of organic chemistry.

Carbon-based fuels, including oil and gas, make up 75 percent of the energy we use. But before these hydrocarbons can be used, the crude oil must be separated into different fractions based on molecular weight. This is carried out in refineries which run day and night. Around 86 percent of the crude oil extracted is used as a fuel, either for heating, for generating electricity or for powering motor vehicles. The remaining 14 percent is used in the manufacture of other chemicals and as a raw material for plastics.

HYDROCARBON CHAINS

ALL organic molecules consist of carbon atoms which each form a total of four covalent bonds. The bonds may be single, double or triple, but they always add up to a total of four. This characteristic makes it possible for carbon-based molecules to form long straight or branched chains with other atoms or to form rings.

The possibilities for different combinations with carbon atoms are almost endless, especially because organic compounds can form isomers – molecules that have the same chemical formula as each other, but have a different structure. Different isomers have different physical properties, and sometimes they can have different chemical properties as well.

To make sense of all this variation, organic chemists group organic molecules containing carbon atoms bonded in the same way into families known as homologous series. All members of a given series have the same general molecular formula and similar chemical properties. However, their physical properties such as melting point, boiling point and density gradually change as the number of carbon atoms increases.

The simplest group of carbon compounds is the aliphatic hydrocarbons, which contain only hydrogen and carbon atoms arranged in straight or branched chains. This group include the alkanes, alkenes and alkynes. But behind their simple formulas lies a great deal of variety.

The alkanes (formerly called paraffins) are made up of hydrogen and carbon atoms joined together by single bonds. The alkane series has the general formula C_nH_{2n+2}, where n represents the number of carbon atoms. It begins with methane (CH_4). Each subsequent compound in the series has one more carbon atom and two more hydrogen atoms than the one before. The next few members, all gases, are ethane (C_2H_6), propane (C_3H_8) and butane (C_4H_{10}). From pentane (C_5H_{12}) onward, the alkanes are liquids. The highest members of the series take the form of waxy solids.

Alkanes are important fuels. They burn cleanly and give off a great deal of energy. The larger alkanes, such as octane (C_8H_{18}), give off more energy per molecule

▷ Alkynes RIGHT are the most reactive hydrocarbon molecules. They contain relatively unstable carbon-carbon triple bonds, so give off a lot of energy when burned. The alkyne ethyne (acetylene, C_2H_2) burns so intensely that it can even burn under water. Alkenes CENTER RIGHT have a double bond between two of their carbon atoms. Because they are unsaturated they are more reactive than alkanes. Alkenes are very important in the manufacture of plastics and polymers, and can also be used as fuels if they are burned in excess oxygen. Alkanes FAR RIGHT are made up of hydrogen and carbon atoms joined by single bonds. Alkanes such as liquid petroleum gas (LPG) and methane are useful fuels because they burn cleanly in the presence of oxygen and give off large amounts of energy.

Ethyne (acetylene)

H —— C ≡≡≡ C —— H

○ Hydrogen

● Carbon

Ethane

$$H - C - C - H$$

△ Natural gas provides clean energy for industrial and domestic uses. Much of the natural gas in Europe is composed principally of the alkane methane (CH_4). Elsewhere, natural gas also contains heavier alkanes such as ethane (C_2H_6) and propane (C_3H_8). When methane is burned in air, carbon dioxide and water are formed, and heat is given out at a rate of roughly 30 kilojoules for each liter of methane burned. Sufficient air is necessary to ensure that methane burns with a clean flame, and to prevent the formation of poisonous carbon monoxide (CO) gas.

△ Polythene ("plastic") bags and polystyrene (styrofoam) – shown in slabs for roadbuilding – are based on the alkene ethene (ethylene, C_2H_4). Ethene is produced by cracking the naphtha and kerosene fractions of crude oil. It is the basis for dry-cleaning fluid and antifreeze.

Ethene (ethylene)

$$H - C - H$$
$$H - C - H$$

than smaller alkanes such as methane because they contain more bonds to be broken.

Alkanes are known as saturated molecules, because it is impossible to add other atoms to them. Saturated molecules are not very reactive, but alkanes can participate in substitution reactions, in which other atoms change places with one or more of the alkane's hydrogen atoms.

Alkenes (formerly called olefins) are hydrocarbons that have a double bond between two of their carbon atoms, with the general formula C_nH_{2n}. Alkenes are unsaturated, and are thus more reactive than the alkanes. They typically participate in addition reactions during which the double bond between the carbon atoms is broken and other atoms are added. Alkenes can be hydrated to form carbohydrates known as alcohols. In this reaction, water reacts with an alkene in the presence of the catalyst sulfuric acid, to form a single-bonded carbon molecule which includes an –OH group.

The smallest alkene molecule, ethene (ethylene, C_2H_4), is probably one of the best known. It is produced industrially during the refining of crude oil, and forms the basis of many familiar plastics, such as "plastic" or polythene (polyethene) bags.

The most reactive hydrocarbons are the alkynes (formerly called acetylenes) They contain a relatively unstable carbon-carbon triple bond, which gives off lots of energy when broken. The smallest molecule in the series, ethyne (acetylene, C_2H_2), is used in acetylene torches for welding because it burns with an intense flame that is hot enough to melt most substances, including metals.

CARBON-HYDROGEN COMPOUNDS

COMPOUNDS that contain hydrogen and carbon provide nearly three-quarters of the energy needs on the planet, and are the most popular form of energy in use today. The energy contained in hydrocarbons was originally trapped from the Sun by ancient plants during photosynthesis that took place millions of years ago. It was converted to chemical energy by plants, and is now preserved in the chemical bonds that hold together the molecules in fossilized hydrocarbon fuels such as oil, gas and coal.

Fossil fuels are formed as the result of millions of years of heat and pressure on the remains of dead plants and animals. After they died, the plants and animals were buried by silt and mud which kept out oxygen and prevented decay. Instead the organic matter was broken down by anaerobic bacteria, which thrive in the absence of oxygen, and gradually became buried by sediments. As the deposits were buried deeper, the pressure increased, and the temperature rose. Over millions of years the material was slowly cooked and converted into long complex chains of hydrogen and carbon.

The type of hydrocarbon that is formed during the long slow "cooking" process depends partly on the chemistry of the original organic material, and partly on the conditions of temperature and pressure under which it was buried. Coal is generally formed as a result of the burial of plants in primeval swamps, and is made up of a mixture of complex hydrocarbons with a high content of carbon.

Both oil and gas are more typically generated by the decay and burial of tiny sea animals such as plankton, although gas can also be generated during the burial of coal swamps. Oil contains less carbon and more hydrogen than coal. Crude oil, pumped from the pore spaces between grains in reservoir rocks buried deep underground, is a mixture of many different kinds of hydrocarbon molecules. Most crude oil is extracted from the ground to make fuel, but around 10 percent is used as a feedstock, or raw material, in the chemical industry. Before it can be used, the various hydrocarbon molecules are separated by refining.

At a refinery, crude oil is separated into different fractions – groups of hydrocarbons that have different boiling points (a function of the number of carbon atoms they contain). The separation takes place in a fractional distillation column, or fractionating tower.

During the refining operation, crude oil heated to a temperature of around 350°C is introduced at the base of the tower. As it boils, the oil vapor passes up the column; as it rises it cools. The different fractions cool and condense at different temperatures, and therefore at different heights along the column. They are first separated and then distilled again to purify them. Light hydrocarbons, which contain between 1 and 4 carbon atoms, and gasoline, which contains between 5 and 10 carbon atoms, condense first near the top of the column. The heavy residues such as bitumen, which contain more than 25 carbon atoms, settle at the bottom of the column.

Complex minerals called zeolites are sometimes used in the refining process to help separate out specific hydrocarbon molecules. Zeolites are ideal for this job because their crystal structure includes pores of specific sizes into which cations (positively charged ions) can fit. Zeolites can be easily modified by replacing some of the atoms in their crystal structure with other elements. This makes it possible for chemists to "design" zeolites that work as very effective molecular sieves for precise purposes, separating out specific types of molecules by size and shape. Alternatively, zeolites can be used as catalysts with molecules of specific sizes.

▲ In most Western countries, gas is transported by pipelines. Gas pipelines are usually buried for safety and convenience, but in cold places such as Alaska, where there is permafrost, the pipelines must remain above ground.

▷ Fossil fuels, such as oil, gas and coal, are formed from the remains of dead plants and animals which have been buried by sediments and "slow cooked" over millions of years in an oxygen-free environment.

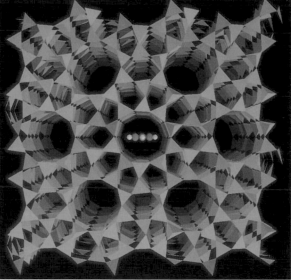

◁ Zeolites, a family of complex aluminum silicate minerals, are sometimes used as molecular sieves to separate out specific hydrocarbon molecules. Although they occur naturally in volcanic rocks, for refinery use they are generally made in the laboratory where it is possible to produce them with precise pore sizes and shapes for separating out specific molecules.

◁ Oil refineries usually have a tall fractionating tower in which separation takes place; storage tanks; and a cracking plant, where heat breaks down large molecules of oil to make smaller molecules for use in gasoline and plastics.

▷ During the fractionation process, crude oil heated to around 350°C is introduced into the base of the column. As it boils, oil vapor passes upward, and cools. Different fractions condense at different heights in the column, where they can be separated. Light fractions include petroleum gas, which condenses at 20°C and is used as bottled gas; gasoline (which condenses at around 70°C); and naphtha, used in chemical manufacture (around 140°C). The middle fractions (190° to 320°C) include kerosene used in jet fuel and paraffin; diesel oil; and fuel oil, used for heating. Heavy fractions, such as lubricating oil and bitumen (asphalt), condense at up to 350°C.

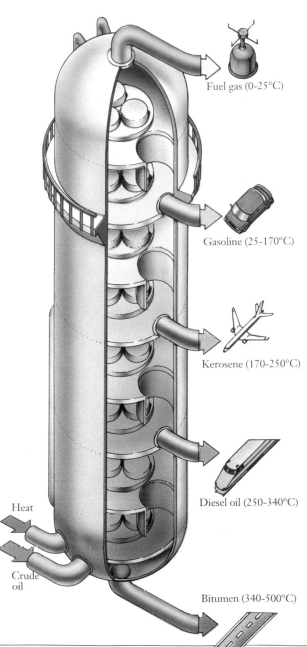

Fuel gas (0-25°C)

Gasoline (25-170°C)

Kerosene (170-250°C)

Diesel oil (250-340°C)

Heat

Crude oil

Bitumen (340-500°C)

CARBON, HYDROGEN AND OXYGEN

A GLASS of wine, a potato and the sugar in coffee all contain compounds of carbon, hydrogen and oxygen. In wine it is alcohol; in potatoes, starch; and the sugar added to coffee is the carbohydrate sucrose. Carbohydrates, like hydrocarbons, can be very large molecules but, unlike hydrocarbons, they also contain oxygen, usually with a ratio of two atoms of hydrogen to one of oxygen. Different functional groups – groups of atoms attached to the hydrocarbon chain – give it the properties that distinguish it from other forms of carbohydrates.

Alcohols – which contain the hydroxyl (–OH) group as a functional group – are used not only for their intoxicating effect. Ethanol (ethyl alcohol, C_2H_5OH), the alcohol in drinks, is used in industry as a solvent for paints, dyes and perfumes; in medicine as an antiseptic and solvent for many drugs; and in some countries as a fuel, because it produces a lot of heat and does not give off polluting sulfur and nitrous oxides.

Ethanol itself is the product of a reaction involving sugar, another carbohydrate. For thousands of years ethanol has been made by fermenting sugar with yeast. All sugars contain the basic units $C_6H_{12}O_6$, but because they differ from each other in the arrangement of the atoms in their molecules, the various sugars have different properties. During fermentation, enzymes in the yeast act as catalysts to break down the sugar into ethanol and carbon dioxide. Ethanol for industrial use is also made by combining a hydrocarbon, the alkene ethene (C_2H_4), with steam (H_2O) to cause an addition reaction in which an –OH functional group is added.

◁ Perfumiers specialize in blending smells characteristic of carbohydrates called esters, which are formed by condensing an acid with an alcohol. Waxes and fats are different forms of esters.

▷ Sucrose ($C_{12}H_{22}O_{11}$) is a carbohydrate produced by sugar cane and sugar beet. It breaks down into fructose and glucose, which have the same formula but different structures.

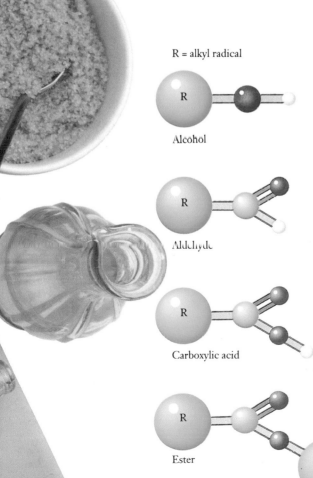

R = alkyl radical

R ——— (Alcohol)

Alcohol

Aldehyde

Carboxylic acid

Ester

◁ **Variations in the formula produce substances as different as sugar, vinegar, potatoes, plastic, and the scent of a flower. Citric acid, which gives lemons and oranges their tangy flavor, is a carboxylic acid; so is ethanoic (acetic) acid, found in vinegar. Carboxylic acids contain a carboxyl (–COOH) group; so do esters, which are used in artificial flavoring and give flowers their sweet smell. Ethanol, the intoxicating alcohol found in wine, contains a hydroxyl (–OH) group. Aldehydes, which contain a carbonyl group (–C=O), are an ingredient in many plastics, including the thermosetting plastic Bakelite, which is a resin made from phenol and methanal. Plastics make good knife handles because they are strong, durable and inexpensive.**

Aldehydes can be made by oxidizing an alcohol (removing hydrogen atoms from it). In aldehydes the functional group is a carbonyl group, –C=O, in which a carbon atom is doubly bonded to an oxygen atom. The carbonyl group is generally linked to two hydrogen atoms or to one hydrogen atom and a hydrocarbon radical. Radicals, which are represented by the letter R, are groups of atoms that attach themselves to different compounds as if they were one element, and remain unchanged internally during chemical reactions.

Aldehydes are easily oxidized and are often used as reducing agents. Formaldehyde (methanal, HCHO), one of the most familiar aldehydes, is a preservative and an important component in plastics. Urea-formaldehyde and melamine-formaldehyde resins are used in paints and lacquers, as adhesives, to add strength to paper products and to make fabrics resistant to creasing.

Alcohol reacts with water and any of the organic acids that contain a carboxyl group (–COOH) to form esters – compounds that give flowers their sweet scent and are the basis of many of the "natural" fruit flavors used in the food processing industry. Esters are also important solvents, widely used in the cosmetic and pharmaceutical industries. Fats are esters with higher molecular weight than the others, and function as food reserves in some plants and animals.

AROMATIC COMPOUNDS

In the aliphatic carbon-based molecules, each carbon atom exhibits its ability to form four covalent bonds with other atoms. This makes it possible for the aliphatic compounds to form straight or branched chains. But in aromatic compounds, carbon exhibits the ability to form strong bonds with itself. Due to this, carbon atoms can form rings as well as in straight and branched chains. Because some natural oils that contain carbon rings smell sweet, carbon-ring compounds were named aromatic organic compounds.

Benzene (C_6H_6), one of the simplest and most important aromatics, is an exception. Its smell is pungent and unpleasant, and there is evidence that it can cause cancer, particularly in young children, so it must be handled with great care. Benzene forms the basis of many compounds, but chemists have only recently understood the nature of the bonding in this relatively common molecule. It consists of six carbon atoms joined in a hexagonal ring, with a hydrogen atom bonded to each carbon. The electrons involved in part of the carbon–carbon bonding are said to be delocalized, and are not associated with any particular carbon atom. As a result, the chemical behavior of benzene can be surprising. For instance, although it is highly unsaturated, and atoms can be added to it during a chemical reaction, benzene is much less reactive than expected, and has its own characteristic properties. Because of the nature of the bonding, benzene tends to undergo reactions that preserve the stable ring.

Nonetheless, benzene rings form the basis of a wide range of aromatic compounds because other atoms or groups of atoms can be substituted into the ring to replace one or more of the hydrogens. Many useful benzene derivatives have been developed, including chemicals used in plastics and dyes (phenylamine or aniline, benzene with an $-NH_2$ group substituted in the ring), rubbers, resins, perfumes (nitrobenzene, benzene with an $-NO_2$ group substituted), preservatives (benzoic acid, benzene with a $-COOH$ group substituted), and flavoring agents (benzaldehyde, benzene with a $-CHO$ group substituted). Substituting a hydroxyl ($-OH$) group into benzene produces phenol, a disinfectant formerly called carbolic acid.

Benzene is also an important molecule in plastics and polymers. Phenol, for example, is used in the

▷ The carbon–carbon bonds in benzene are unique – they are neither double nor single bonds. The benzene ring is a flat hexagon with six C–C bonds all of the same length. The bond lengths are less than for a C–C single bond, but greater than for a C=C double bond. In benzene, some of the electrons involved in the bonding are said to be delocalized. They are not associated with any particular carbon atom, but are shared by all six carbon atoms in the ring. There is a uniform electron density in all the C–C bonds. A benzene ring is shown here viewed from three different perspectives: edge-on, tipped and from on top. The cloud of delocalized electrons is clearly visible. Because of the unique nature of the bonding in benzene rings, the chemistry of benzene holds many surprises.

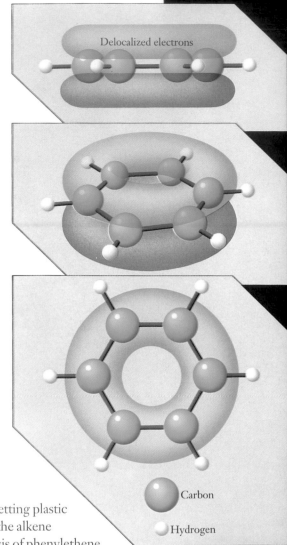

Delocalized electrons

Carbon

Hydrogen

manufacture of the thermosetting plastic Bakelite. Benzene added to the alkene ethene (C_2H_4) forms the basis of phenylethene (styrene), the building block of the polymer polystyrene. This important synthetic polymer is made up of benzene rings arranged along the length of a chain of carbon atoms.

There are three different types of polystyrene: an amorphous form in which the arrangement of benzene rings is random; and two crystalline forms, one with benzene rings on just one side of the chain, and another in which they alternate on either side of the chain. Polystyrene can be used to make molded objects and electrical insulators. Because it is clear and can be colored easily, it is often used for optical components.

Benzene rings can also join together to give fused ring systems which, like benzene itself, can participate in addition and substitution reactions to form a wide range of compounds. In these systems the delocalization of the electrons extends over all the rings. Compounds of this type include naphthalene, which consists of two benzene rings fused together. Naphthalene is used in plasticizers, alkyd resins, polyesters, insecticides and moth balls.

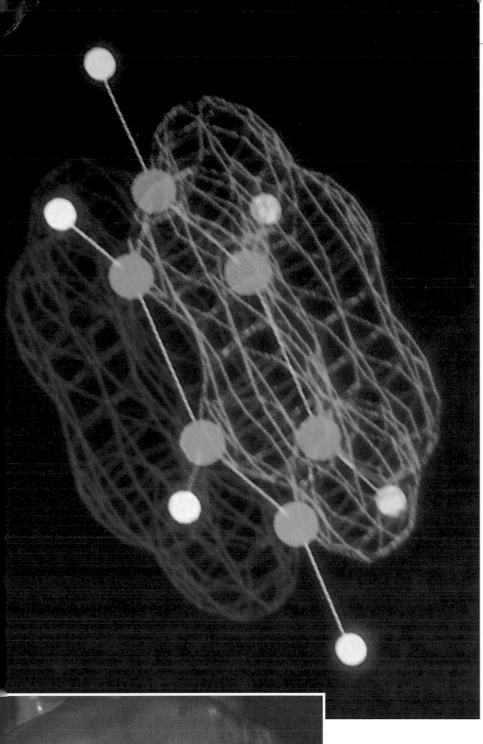

▷ There are many different ways of representing the benzene structure. They include a simple hexagon 1; a hexagon with bonds to which substituents can be attached 2; a hexagon with alternate double and single bonds 3 (according to the model proposed by the German chemist August Kekulé in 1856); and a hexagon containing a circle containing a circle 4. Benzene rings can be combined to form other compounds. For example, two fused benzene rings make up naphthalene, an important compound used in the manufacture of plasticizers, resins and polymers. Anthracene, a component of coal tar, is made up of three benzene rings fused in a row. Three benzene rings fused in the shape of a dog's leg comprise phenanthrene, a compound which is a component of steroid hormones.

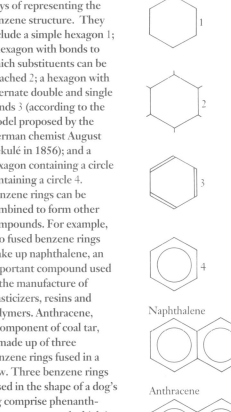

Naphthalene

Anthracene

Phenanthrene

Benzapyrene

CARBON RINGS AND STEROIDS

Steroids widely known as muscle-building drugs used illegally by athletes – are organic compounds whose central molecule contains four carbon rings (a sterol), sometimes with side-chains attached. Steroids include a wide variety of compounds: human sex hormones and bile salts, plant alkaloids such as caffeine and nicotine; and the insect hormone ecydsone, which controls molting. All steroid hormones are based on cholesterol, which is made up of a five-sided carbon ring attached to phenanthrene, a compound of three benzene rings. There is very little chemical difference between the male sex hormone testosterone and the female hormone progesterone which contains one more oxygen atom.

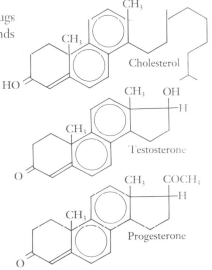

Cholesterol

Testosterone

Progesterone

MAKING HYDROCARBONS

ADDING or subtracting hydrogen atoms can make a big difference to the preparation of edible oils and fats. At room temperature, oils are liquids, whereas fats are solids. The physical difference between an edible oil, such as olive oil or sunflower oil, and a solid fat, like margarine, is largely a result of differences in the number of hydrogen atoms they contain.

Edible oils and fats have a similar chemical structure. Both are made up of long chains of carbon atoms with hydrogen atoms attached. Oils, however, are more unsaturated than fats – that is, they have a higher proportion of carbon–carbon double bonds, and as a result, fewer hydrogen atoms. Like all unsaturated compounds, oils are more reactive. The double C=C bonds are less stable than single bonds and help to give oils a lower melting point. This is why oils are liquids at room temperature. Fats, on the other hand, are said to be saturated: all their carbon atoms are involved in four single bonds, and thus are not looking to form new bonds. As a result, they have a higher melting point, and are solids at room temperature.

Adding hydrogen to oils to change them to fats is called hydrogenation. The oil is first heated to 150°C and a powdered nickel catalyst is added. When hydrogen is bubbled through the oil it causes the double C=C bonds in the carbon chain to break. An addition reaction takes place when hydrogen is added to the chain as the carbon atoms begin to form new single bonds with the hydrogen.

The more hydrogen that is used, the more double bonds are broken. To produce a hard fat, with a relatively high melting point, hydrogen is added until the oil is fully saturated and there are no more double bonds to break. To produce a softer fat, with a lower melting point, less hydrogen is added. As a result, the oil does not become fully saturated – a few double bonds still remain.

Unsaturated fats, such as soft margarines, are "high in polyunsaturates" – made up of fats and oils that contain many double bonds. These are easily spread, and many doctors believe that they are healthier than saturated fats, which may cause blockage of the blood vessels and lead to heart disease.

Margarine is an emulsion that contains droplets of water, skim milk and brine, suspended in oil and fat.

KEYWORDS

DOUBLE BOND

EMULSIFIER

FAT

HYDROGENATION

OIL

SATURATED COMPOUND

SINGLE BOND

UNSATURATED COMPOUND

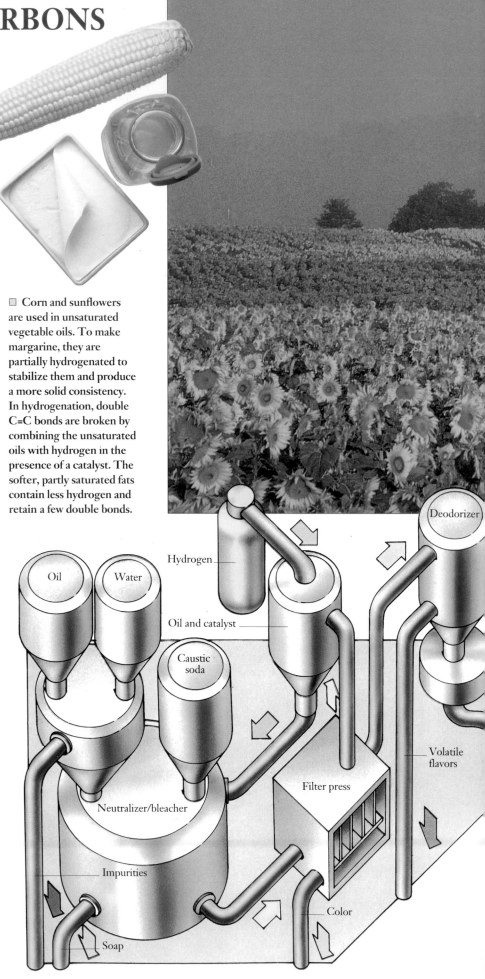

☐ Corn and sunflowers are used in unsaturated vegetable oils. To make margarine, they are partially hydrogenated to stabilize them and produce a more solid consistency. In hydrogenation, double C=C bonds are broken by combining the unsaturated oils with hydrogen in the presence of a catalyst. The softer, partly saturated fats contain less hydrogen and retain a few double bonds.

Oil

Water

Hydrogen

Oil and catalyst

Caustic soda

Deodorizer

Volatile flavors

Filter press

Neutralizer/bleacher

Impurities

Color

Soap

Non-margarine low-fat spreads contain a higher proportion of water. An emulsifier is used to make the oil and water mix in all types of margarines. In the low-fat versions, extra emulsifiers and stabilizers must be added; these may be ingredients such as milk solids, the milk protein casein, vegetable proteins and vegetable gums.

Edible oils are not the only substances that can be hydrogenated. Coal can be hydrogenated by a process called liquefaction to produce a liquid fuel, which can then be refined to produce gasoline, heating oil, diesel fuel, jet fuel, fuel oils and petrochemicals. When coal is liquefied, the ratio of hydrogen to carbon atoms is greatly increased. In direct liquefaction process, called the Bergius process, pulverized coal is suspended in a liquid mixture of hydrocarbons. The slurry is then hydrogenated by being heated and exposed to gaseous hydrogen under pressure. The liquids produced as a result of the process are first separated from the ash and then distilled to obtain different hydrocarbon fractions.

In the Fischer-Tropsch process, hydrogen and carbon monoxide are heated to 200°C and passed over a catalyst of cobalt or nickel. This yields a mixture of hydrocarbons which can be separated into various fuels by distillation. Alcohols and ketones are also produced during the hydrogenation process.

◁ **To make margarine, oils, which today generally consist of vegetable oils, are first treated to remove impurities such as carbohydrates, lipids and resins. The oil blend is then mixed with a solution of milk, salt and emulsifiers to form an emulsion. The milk may be cultured first with strains of bacteria to convert the milk** sugar lactose to lactic and other short-chain organic acids. Afterwards, the emulsion is pumped through a chilling unit, known as a votator, where it is cooled quickly. The soupy liquid which remains is allowed to rest in a second votator until firm. Rotating blades scrape the margarine off the walls of the vessel and whip it before it is moved to a tempering tube for extrusion and packing. In some processes, vitamin A – found naturally in butter – is added. The color of the product is adjusted using artificial coloring agents. "Low-fat" margarines contain added water.

POLYMERS
and Plastics

4

ALL LIVING THINGS contain polymers: proteins, carbohydrates, wood and natural rubber are all polymers. What nature has invented, chemists have learned to copy and manipulate successfully. It is now possible to make a vast range of synthetic polymers – from plastics to synthetic fibers – with properties to suit specific needs.

Polymers are large organic macromolecules that usually contain carbon and hydrogen, often along with oxygen and nitrogen. They are made up of smaller repeating units known as monomers. Some polymers, called homopolymers, contain just one monomer; but others, known as copolymers, are made up of two or more types of monomers. During polymerization, the monomers join up to make long chains. Although polymers can contain as few as five repeating units, most are made up of several thousand such units. Polymer chains, which can occur as straight chains or branched ones, often contain more than 50,000 carbon atoms.

The nature of the groups of atoms that make up a polymer determines the characteristics of the polymer. The molecules in the chain can form bonds with each other, both along their length or across chains. These intermolecular forces give different plastics and synthetic fibers their individual characteristics.

Plastic used to be associated with cheap, poor-quality goods, but modern high-tech polymers are known for their reliability and chosen for their suitability for the job at hand. The use of plastic has revolutionized sports, where lightweight, strong plastics replace natural materials in many types of equipment. The paraglider shown here trusts his life to the superior properties of plastics and polymers. His canopy is made out of nylon, a strong polymer fiber. Another polymer fiber is used in the ropes, and a rigid, tough plastic helmet offers lightweight protection for his head.

NATURAL POLYMERS

CHEMISTRY imitates nature, and polymer chemists have learned as much from studying natural materials as they have from making products in a test tube. Natural polymers are everywhere – in fact, many of them grow on trees.

The sap obtained from rubber trees provided the raw material for one of the earliest polymer industries: rubber production. Natural rubber, or latex, known to chemists as polyisoprene, is a polymer made up of 1000–5000 monomers of the unsaturated hydrocarbon isoprene (C_5H_8), which is obtained from the sap of the rubber tree. Isoprene contains two double bonds separated by a single bond. When it polymerizes, bonds are broken and rearranged to allow the monomers to link up in a long coiled chain. This chain gives rubber its physical properties.

Rubber is elastic because stretching tends to straighten out the entangled chains, but when the stretching force is released, the intermolecular forces pull the chains back together. The tangled chains are also what gives rubber its ability to hold together when stretched, rather than to break or crumble. The fact that rubber is made up of a dense tangle of hydrophobic hydrocarbon chains also accounts for its waterproof properties.

Natural rubber softens on heating and hardens on cooling without changing its chemical properties. It is used in cements and adhesives, and as tape for wrapping cables and insulating electrical equipment. However, because it has a relatively low melting point, natural rubber softens in warm weather.

The process of vulcanization is used to raise rubber's low melting point, and to make it hard enough for use in tires and other products. During vulcanization, the long chains of polyisoprene units are linked together by sulfur bonds, often called sulfur bridges or cross-links. This cross-linking makes natural rubber more thermally stable, but changes it into a material that cannot be altered once molded or formed without destroying the chemical properties of the product.

A further drawback is that vulcanized rubber has a relatively short lifetime, because the sulfur bridges react readily with the oxygen in air. As a result, natural rubber supplies less than 40 percent of the world's needs, and is largely being replaced by synthetic rubber substitutes, such as nitrile and chloroprene rubbers produced from petroleum. These synthetic

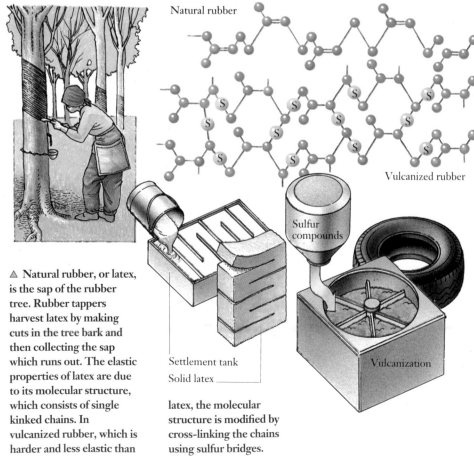

Natural rubber

Vulcanized rubber

Sulfur compounds

Settlement tank
Solid latex

Vulcanization

▲ Natural rubber, or latex, is the sap of the rubber tree. Rubber tappers harvest latex by making cuts in the tree bark and then collecting the sap which runs out. The elastic properties of latex are due to its molecular structure, which consists of single kinked chains. In vulcanized rubber, which is harder and less elastic than latex, the molecular structure is modified by cross-linking the chains using sulfur bridges.

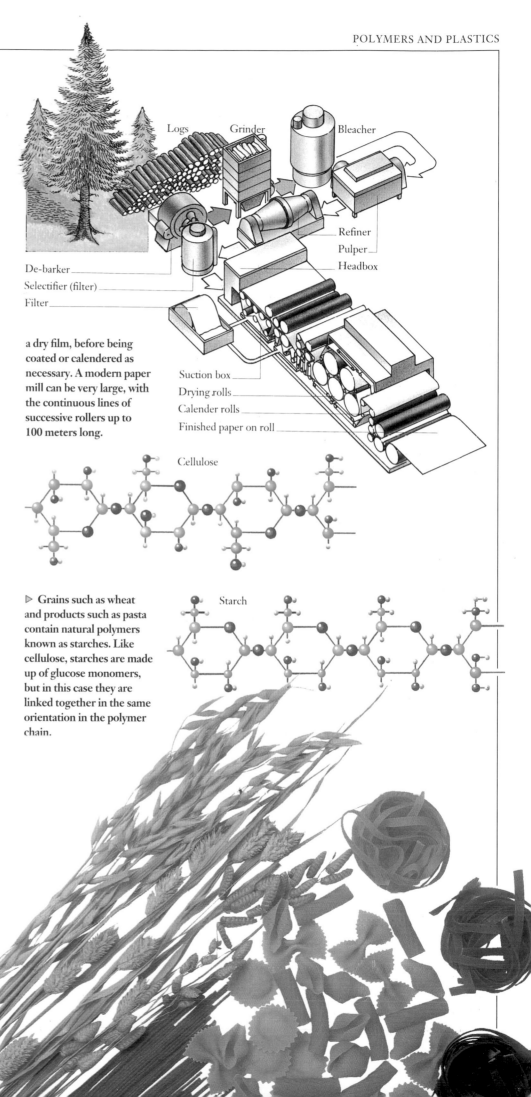

◁ **Cotton, shown here being harvested, comprises fibers of cellulose, the structural ingredient in most plant cells. The orientation of the glucose monomers in the cellulose poylmer alternates, resulting in strong fibers.**

▷ **Cellulose fibers are used to make paper and textiles. The process of paper-making itself is simple, but due to requirements for quality and pollution control, it has become more complex. In paper-making, de-barked logs, sometimes mixed with cotton rag or other fibers, are pulped mechanically or chemically, then bleached, to make a slurry of fibers and water. This slurry is poured over a fine mesh in huge papermaking machines to form large flat sheets. The water is removed by rolling to leave** a dry film, before being coated or calendered as necessary. A modern paper mill can be very large, with the continuous lines of successive rollers up to 100 meters long.

Logs   Grinder   Bleacher

De-barker
Selectifier (filter)
Filter
Refiner
Pulper
Headbox

Suction box
Drying rolls
Calender rolls
Finished paper on roll

Cellulose

▷ **Grains such as wheat and products such as pasta contain natural polymers known as starches. Like cellulose, starches are made up of glucose monomers, but in this case they are linked together in the same orientation in the polymer chain.**

Starch

versions have higher melting points and are far less affected by organic solvents.

Cellulose and starch are other important natural polymers that grow on trees. Both are based on the monomer glucose, a simple sugar formed by plants during photosynthesis. Starch is used by both plants and animals as an energy store. When energy is needed the starch can be easily broken down into glucose molecules, which are then oxidized to release carbon dioxide, water and energy. Cellulose is the polymer that makes up the main structural material of plants. It is one of the most abundant organic substances on Earth, and cannot be readily broken down; herbivorous animals that depend on eating cellulose often have special bacteria living in the gut to assist with digestion.

Chemists can combine a polymer with another material to produce a composite with special properties, often for engineering applications. Fiberglass is a composite in which thin glass fibers are bound together in an unsaturated polyester resin. This process occurs in nature too. One of the most abundant natural polymer composites is a cellulose fiber-reinforced phenylic resin composite – best known under its common name, wood.

SYNTHETIC POLYMERS

IMITATING nature was the goal of the first polymer chemists, and polymer science evolved as they found ways of synthesizing materials with properties similar to those already available in the natural world. Now polymer chemists use their knowledge to exploit the possibilities of designing molecules with properties to order. Initially they concentrated on trying to develop less expensive or more readily available substances as alternatives for natural materials, such as synthetic wood and rubber for use in construction, synthetic cotton and silk to make into fabrics, and synthetic resins for use in paints and varnishes.

Some of the earliest commercial polymers were based on naturally occurring polymers, such as cellulose. Cellulose is an important structural material in plant cells, and is the major component of wood and cotton. It also forms the basis of one of the first synthetic fibers to be developed, artificial silk, or rayon.

In rayon, the protein fibers that make up natural silk and give it its shiny appearance are replaced by synthetic fibers. These are made by treating cellulose from wood pulp or cotton with an acid. In the first form of rayon, the cellulose, mostly derived from purified cotton fibers, was reacted with nitric acid to form cellulose nitrate ("nitrocellulose"), which could be formed into fibers. However, because molecules containing nitro groups are dangerously flammable and even explosive, the cellulose nitrate form of rayon was not ideal for use in clothing.

Today's rayons are much safer. They are produced by reacting cellulose with acetic acid to produce cellulose acetate. This is extruded through spinnerets to form fibers, which can be spun into a yarn and woven to produce the textile rayon acetate.

Synthetic rubbers were developed to find ways of taking advantage of some of the useful properties of natural latex, such as its elasticity and waterproof nature, while at the same time overcoming some of its drawbacks, such as low melting point and relative instability in air. In making synthetic rubber, chemists aimed to replace the isoprene monomers with other hydrocarbons derived from crude oil.

Synthetic rubbers can be made using either emulsion or solution polymerization. In emulsion polymerization, the reaction is carried out in water in the presence of a catalyst, and an emulsion of rubber is formed. In the process of solution polymerization, the reaction takes place in an organic solvent.

Today roughly 60 percent of rubber is produced synthetically. By choosing monomers carefully it is possible to make synthetic rubbers with specific properties to suit different applications, and with improved properties of toughness and durability. One of the major rubbers produced today is a copolymer of styrene (C_8H_8) and butadiene (C_4H_6), that is widely used for vehicle tires. Neoprene, an oil-resistant rubber which is not affected by ozone, is made by the polymerization of chloroprene (C_4H_5Cl).

Hydrocarbons are also key raw materials in the production of plastics. Nearly 4 percent of crude oil production is used as a raw material for plastics. Crude oil includes a variety of hydrocarbon molecules which contain varying numbers of carbon and hydrogen atoms. The naphtha fraction (8 to 12 carbon atoms) is the most important for plastics production because it can be "cracked", or broken down into smaller molecules, by passing its vapor through tubes heated to about 800° C and mixing it with steam. One of the products of the cracking of naphtha is the alkene ethene (ethylene, C_2H_4) – the single most important substance for producing plastics.

▷ Ethene (ethylene, C_2H_4) forms the basis of many plastics and polymers. Butadiene, the synthetic rubber molecule, resembles two joined ethene molecules. When one of the hydrogen atoms in ethene is replaced by chlorine, the result is chloroethene (vinyl chloride), the monomer at the heart of the common plastic PVC.

The monomer butadiene ($CH_2:CH.CH:CH_2$) is used to make synthetic rubber, based on poly-butadiene. Like natural rubber, the synthetics can be vulcanized and made into tires for aircraft and automobiles.

Vinyl chloride (chloro-ethene, $CH_2:CHCl$) polymerizes to give polyvinyl chloride, used to insulate electric cables and for audio disks, waterproof clothing and the interior trim for cars.

Styrene (phenylethene, $CH_2:CH(C_6H_5)$) polymerizes to form polystyrene, a thermo-plastic is used in ballpoint pens, injection-molded plastic toys and models. Styrofoam, a rigid foam, is used for drinking cups.

◁ Synthetic rubbers, such as polybutadiene (made by dehydrogenation of butane and butene), are a reliable material for use in high-stress conditions. Here, smooth tires made of soft synthetic rubber provide a motorcycle racer with the grip he needs to move safely on a dry track.

▽ Viscose rayon, cellophane and cellulose sponges are made from regenerated cellulose. The viscose solution is mixed with suitable fibers and salt crystals, and then acidifying and leaching the mixture in water to form a sponge. To make rayon fibers, the viscose solution is forced into an acid solution through tiny holes in a spinneret. In cellophane manufacture, the viscose is squeezed through a narrow slot into an acid bath.

▽ The development of a process to manufacture viscose rayon, a type of artificial silk, was one of the earliest success stories in polymer chemistry. Like cellophane and cellulose sponges, viscose rayon is made from regenerated cellulose, which is generally obtained from wood, wood pulp or cotton. In the viscose process, purified cellulose is treated with sodium hydroxide (NaOH) before being shredded and aged. It is then reacted with carbon disulfide (CS_2) and sodium hydroxide (NaOH) to form a viscose solution.

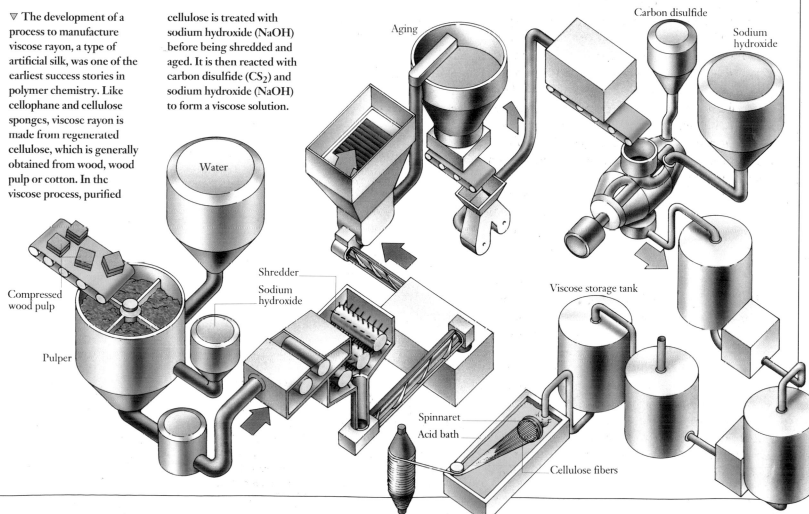

Aging

Carbon disulfide

Sodium hydroxide

Water

Viscose storage tank

Compressed wood pulp

Pulper

Shredder

Sodium hydroxide

Spinnaret

Acid bath

Cellulose fibers

TYPES OF PLASTIC

PLASTICS have a wide range of physical properties. Some, like Bakelite, are hard and will not melt. Others, such as polythene, can be made into thin flexible films. One of the advantages of plastics is that they can be synthesized to have different properties.

The physical properties of plastics are closely related to the molecular structure of the polymer chains and the nature of the forces that link the chains together. Thermoplastics such as polythene, polystyrene (styrofoam), polyvinyl chloride (PVC) and nylon, which melt on heating, are made up of long, thin, covalently bonded monomers that form tangled chains. Relatively weak electrostatic forces, including induced dipole forces, dipole-dipole forces and hydrogen bonding, hold the chains together. These forces can easily be destroyed by heat: the plastics soften on warming as the chains begin to move over each other more easily. This is why thermoplastic polymers stretch and flex easily, melt at low temperatures and melt without decomposing.

In contrast, thermosetting polymers, such as Bakelite and other methanal (formaldehyde) resins, including urea-formaldehyde and melamine, are made up of cross-linked chains. Strong covalent bonds both within and between the chains form a random three-dimensional network with rigid bonds, which inhibit the chains from moving in relation to one another when they are heated, stretched or compressed. The covalent bonds between the chains develop as the thermosetting polymers are cured (usually by heating) after they are molded or shaped. Once the bonds form between the polymer chains, the plastics become rigid and hard. Their shape cannot be changed and they burn or char before they melt.

Chain length and branching in the chain also affect physical properties. Increasing the length of the polymer chain tends to produce stronger materials, because longer chains tangle more easily, and thus have more points of contact with neighboring chains. This leads to a greater number of attractive forces between molecules which hold the chains together. As a result, when one molecule in the chain moves, it pulls other molecules with it, creating a stronger material.

Straight chains can pack together more closely to form high-density

polymers that are strong but not very flexible, and that soften at relatively high temperatures. Polymers made of highly branched chains that cannot pack together very closely tend to be less dense, and may be glassy and transparent.

The nature of the bonding within the chain is another important variable. Polymers normally act as electrical insulators, but in 1976 conducting polymers were developed. These have alternating single and double carbon-carbon bonds in their polymer chains and become electrically conducting when "doped", or treated with chemicals that either donate or remove electrons to leave them partly reduced or oxidized.

Conducting polymers are already used by two Japanese companies to make small button batteries for cameras and hearing aids. Now research chemists are working to develop polymer blends, or mixtures of conducting and non-conducting polymers. These will result in a material with electrical properties that can be used like normal plastics to produce a wide range of products. Plastics that shield against electromagnetic radiation, plastic electrodes, conducting paints, inks, fibers and flooring materials are just some of the possibilities.

▷ Oil, natural gas, limestone, salt and fluorspar are major raw materials in the production of monomers – molecules which are linked together to form plastics and polymers. The physical properties of polymers can be controlled by changing the types or arrangement of monomers in the chain. There are two main types of plastics: thermosets, such as Bakelite, which cannot be softened once they have been cooled; and thermoplastics, such as polyethylene and polystyrene, which can be softened by warming and then shaped.

▽ Polyethylene BOTTOM and nylon BELOW are two of the most familiar straight-chain polymers. Polyethylene, the material used for plastic bags, consists of chains made up only of carbon and hydrogen; nylon has additional molecules of nitrogen and oxygen. The bonds that hold the chains together are easily destroyed by heating.

Nylon 66

Polyethylene

Source	Raw mate...
Crude Oil	Acetylene
	Benzene
	Butadiene
	Ethylene
	Propylene
Natural Gas	Carbon monoxi...
	Hydrogen
Air	Nitrogen
Salt	Chlorine
Limestone	Calcium cyanan...

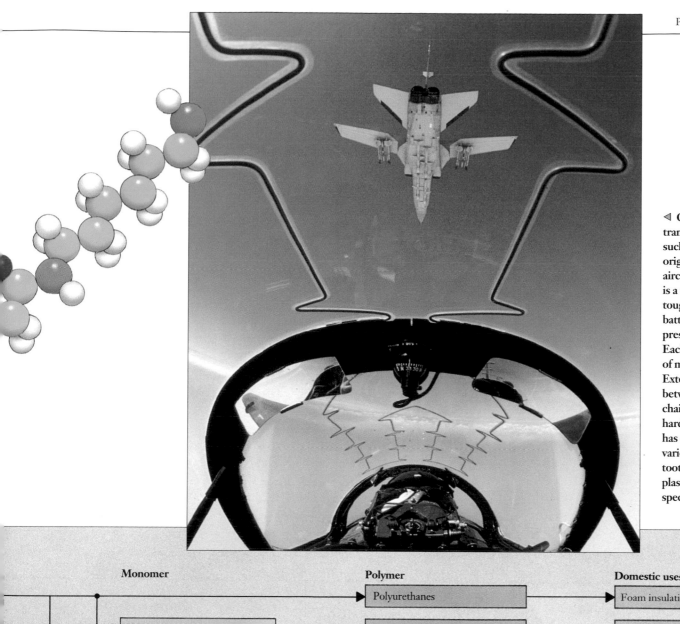

◁ Optically clear, transparent acrylic plastics, such as Plexiglass, were originally developed for aircraft canopies. Plexiglass is a thermoset plastic, tough enough to withstand battering by wind and air pressure at high altitudes. Each molecule is made up of millions of atoms. Extensive cross-linking between the polymer chains makes the plastic hard and tough. Plexiglass has other uses in a wide variety of objects from toothbrushes to false teeth, plastic rulers, lenses and spectacle frames.

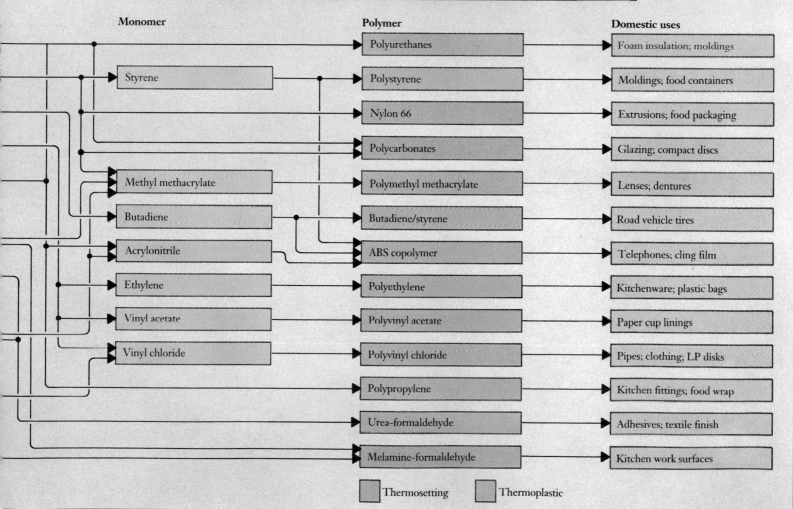

Monomer	Polymer	Domestic uses
	Polyurethanes	Foam insulation; moldings
Styrene	Polystyrene	Moldings; food containers
	Nylon 66	Extrusions; food packaging
	Polycarbonates	Glazing; compact discs
Methyl methacrylate	Polymethyl methacrylate	Lenses; dentures
Butadiene	Butadiene/styrene	Road vehicle tires
Acrylonitrile	ABS copolymer	Telephones; cling film
Ethylene	Polyethylene	Kitchenware; plastic bags
Vinyl acetate	Polyvinyl acetate	Paper cup linings
Vinyl chloride	Polyvinyl chloride	Pipes; clothing; LP disks
	Polypropylene	Kitchen fittings; food wrap
	Urea-formaldehyde	Adhesives; textile finish
	Melamine-formaldehyde	Kitchen work surfaces

☐ Thermosetting   ☐ Thermoplastic

SHAPING PLASTICS

Nylon can be molded to make syringes or spectacle frames, or drawn into fibers to make stockings. Once polymers are synthesized, they can be processed in many different ways to provide a wide variety of shapes and structures suitable for many uses.

Fibers can be made by cold drawing, in which a liquid polymer is stretched into a long thin filament. During drawing, a thin neck forms in which the polymer chains become oriented along the length of the fiber and form a more crystalline structure. This leads to increased strength. Cold drawing produces a fiber that is pliable, elastic, flexible and tough.

Fibers can also be produced by extrusion, in which small granules of polymer are melted and forced through a nozzle. The result is a continuous length of fiber, the same shape all the way through. Dacron (Terylene) fibers can be produced in this way.

Extrusion is also used to produce sheets, films and various types of tubes and pipes. In extrusion, the thermoplastic raw material in powder or granule form is fed into a heated cylinder. A screw mechanism or hydraulic plunger in the cylinder pushes the material forward and compacts, melts and homogenizes it. The plastic mass is shaped by being forced through a die as it emerges from the cylinder. Extruded objects usually require further processing before they can be considered finished goods.

Injection molding can be used to produce a wide variety of shapes with great accuracy. In an injection molding machine an injection unit, consisting of an extruder with a screw that can be moved backward and forward, melts the feed material as the screw moves backward and forces the material out into the mold as the screw is driven forward. The mold is opened to remove the finished item. Goods produced in this way generally do not need further processing.

To make hollow objects, such as bottles, cans or other containers, extrusion blow molding is used. In this process, an extruder forces a plastic tube vertically downward between the two halves of an open two-part mold. After the mold is closed and sealed, compressed air is blown into the tube to force the soft plastic against the sides of the mold and form the desired hollow shape.

Calendering is used to produce semifinished goods, such as coverings, flooring and plastic sheets for use as

▷ Vacuum forming is used to form thin sheets of thermoplastics into simple shapes. The first step is to make a mold of the desired shape. A heated sheet of plastic is placed over the mold, and the air removed to force the plastic to take on the shape of the mold. The process is relatively slow and difficult to automate. Vacuum forming is commonly used to make novelty items, such as these hats. It is also useful for making items such as chocolate box liners or other types of packaging which mirror the shape of the goods.

◁ Polyethylene film is shown here being formed into a continuous tube, by an extrusion process, for subsequent commercial use as shopping bags.

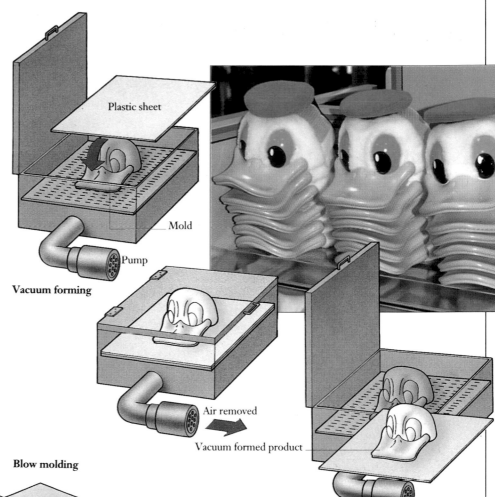

Vacuum forming

Plastic sheet

Mold

Pump

Air removed

Vacuum formed product

Blow molding

Hot air

Metal mold

Thermoplastic tube

Molded product

△ Blow molding is a fast and efficient technique used to make hollow objects such as bottles and containers out of thermoplastics. It is easily adapted to continuous production. To make a hollow object, a tube of plastic is placed inside an open two-part hinged mold. The mold is closed and heated to soften the plastic, which is then pressed against the sides of the mold using air pressure. The finished object is removed by opening the mold, and any excess plastic is trimmed off.

wrappings. Polyvinyl chloride is often processed in this way to produce PVC sheeting and to coat fabrics. Calendering machines work on a similar principle to old-fashioned kitchen mangles. A melted mass of polymer is passed over and through a series of heated rollers to form a sheet. The sheets can be further processed by, for example, stamping, to produce finished goods.

Many polymers, such as polystyrene, polyethylene and PVC can be expanded or foamed to reduce their density. In foaming, a gas is included in the polymer structure. This can be done by mixing compressed air or gas into the melted polymer mass. As processing goes on, the gas is released, expands, and forms a blowing agent to foam the polymer. Another method involves adding a blowing agent such as sodium bicarbonate to the plastic raw materials. When the hot melt is formed, the sodium bicarbonate decomposes to give off carbon dioxide gas which forms bubbles in the foam. As the polymer cools and solidifies, the bubbles become fixed in the structure. Expanded polymers can be made into products ranging from egg boxes and drinking cups to sponges and steering wheels for cars by methods such as injection molding, extrusion and calendering.

USING PLASTICS

A WIDE range of properties can be designed into polymers, with the result that polymers are in use almost everywhere. Tough plastics are replacing ceramics and cast iron. Polymer-based synthetic fibers are replacing wool and silk to make strong carpets, and replacing other natural fibers to make clothing that lasts longer and needs no ironing.

Major differences in the physical properties of polymers result from the types of intermolecular forces between chains, and account for the great differences between hard rigid thermosetting plastics and thermoplastic polymers, which can be shaped again and again because they become soft and moldable when heated without undergoing any chemical change.

One of the best known thermosetting plastics, and the first to be invented, is Bakelite. It is produced using a condensation reaction involving phenol (C_6H_5OH) and methanal (formaldehyde, HCHO). During curing at high temperatures, strong covalent bonds form between the polymer chains and result in a strong rigid material. Bakelite is a good electrical insulator and can be machined and dyed. It is still commonly used for distributor caps on cars.

Differences in the nature of the polymer chains themselves also play an important role in controlling physical properties. Differing arrangements of the monomer propene (C_3H_6) in a polymer chain are responsible for the differing physical properties in the three forms of the familiar polymer polypropylene. In one form, the polymer chain contains monomers arranged in a regular sequence. This results in a polymer with good mechanical strength. In a second form, the chains are compressed to form a crystalline polymer which is used to make blow-molded items such as bottles, and for packaging film. In atactic polypropylene, the chains are disordered. This results in a pliable material that is slightly tacky to the touch. It is used in backings for carpet tiles, and mixed with bitumen for use on roads and roofs.

Branching in the polymer chain is another important variable. The familiar thermoplastic polymer polyethene (polyethylene, best known as polythene) is available in low- and high-density forms whose differing physical properties relate to the amount of branching in the polymer chain. Low-density polyethylene (LDPE), which is formed during a high-pressure polymerization process, is made up of highly branched chains. In contrast, the polymer chains in high-density polyethylene (HDPE), produced using a low-pressure polymerization process, are generally linear and contain few branches. The branching chains of LDPE cannot be tightly packed. As a result, LDPE is relatively soft and has a low level of crystallinity and density. The linear HDPE chains can be closely packed. HDPE is harder than LDPE, has a higher crystallinity and higher density. It is stronger and less easily deformed by heat than LDPE, and can be easily molded into complicated shapes. It is often used to make automobile fuel tanks, water tanks and piping.

Polymers can also be used to form fibers that can be woven or knitted to form a fabric. The first totally synthetic fiber was nylon, a polyamide incorporating

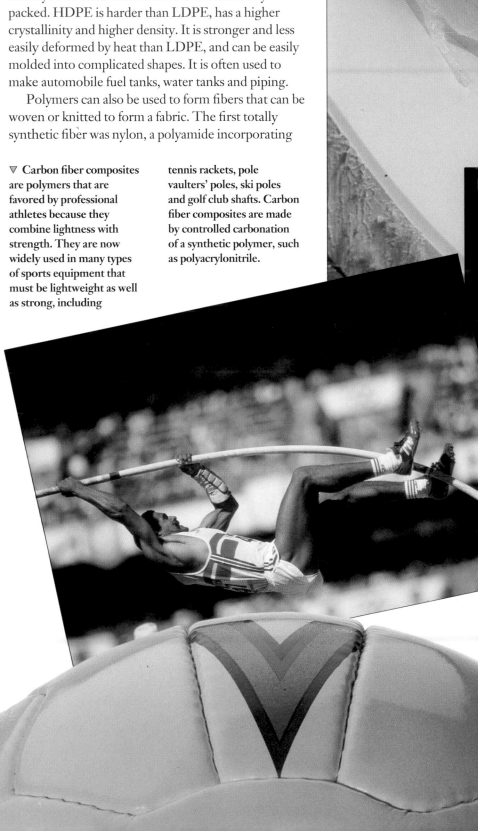

▽ **Carbon fiber composites are polymers that are favored by professional athletes because they combine lightness with strength. They are now widely used in many types of sports equipment that must be lightweight as well as strong, including** tennis rackets, pole vaulters' poles, ski poles and golf club shafts. Carbon fiber composites are made by controlled carbonation of a synthetic polymer, such as polyacrylonitrile.

◁ Fiberglass is a popular material for making boats. Here, glass cloth is being applied to the mold of the hull of a boat. In a separate operation, resin will be added to complete the composite. Fiberglass is made by painting an epoxy resin onto a glass fiber cloth or mat to make a stiff, strong, yet lightweight composite material. The resin is applied as a liquid, which hardens on drying to provide stiffness. The strength in fiberglass comes from the glass fibers.

◁ Polymers can also be used to provide all-round protection. A hard plastic helmet, made of reinforced polycarbonate, protects the driver's head. Her suit is made of the polyurethane-based fiber lycra, which is warm and lightweight.

▽ One of the triumphs of polymer chemistry is in the imitation of natural materials such as leather. Here, vinyl is used instead of leather in a soccer ball. Vinyl chloride ($H_2C=CHCl$) is the starting material for polyvinyl chloride, or PVC, one of the most successful and familiar plastics.

the monomers diaminohexane and adipic acid. Nylon is characterized by high strength, elasticity, toughness, abrasive resistance and low-temperature flexibility. It is also generally resistant to most solvents, acids, bases and outdoor weathering.

Nylon fibers are produced by cold drawing, a process that causes all the polymer chains to become oriented along the length of the fiber. Nylon fibers are widely used in textiles, including stockings, as well as for tire cord, rope, threads and belts.

Cold drawing also provided the breakthrough needed to produce polyester fibers from polyester, a polymer formed by a condensation reaction between polyhydric alcohols and organic acids. Polyester fibers are strong and resistant to heat, and when mixed with cotton fibers they make cool, comfortable clothing that does not crease easily.

Amides, the monomer units in nylon, provided the basis for the development of polyaramids, which make up fire-resistant materials such as Kevlar. This is a low-density fiber, made from carbon, hydrogen, oxygen and nitrogen – all light atoms. Yet, weight for weight, it is nearly five times as strong as steel. The strength is derived from the arrangement of the polymer chains which line up parallel to one another and are held together by hydrogen bonds to form sheets of molecules. The sheets are stacked together regularly around the fiber axis to give a very well-ordered structure.

Tailoring polymers for specific purposes can be taken even further in composite materials, which normally consist of two phases – a continuous matrix and a fibrous reinforcement. The reinforcement contributes strength and stiffness, while the matrix binds the fibers together to produce a consolidated structure which allows stress to be transferred from one fiber to another. Careful choice of the matrix can produce some useful engineering properties. The glass fiber and polypropylene composite RTC can be shaped by either low-temperature, low-pressure molding to make objects such as interior panels for cars, or by high-pressure flow molding to produce complex rigid shapes.

RECYCLING POLYMERS

PLASTIC rubbish is a common but unwelcome sight around the world. Even the open ocean is not free of it, and plastic litter from ships, yachts and oil rigs spoils the beauty of beaches around the world. Many people question the merits of using plastics in the first place. Often plastics are considered undesirable when compared to more "natural" materials such as paper. However, careful analysis of the environmental cost of using plastic versus paper carrier bags, carried out by the German Federal Office of the Environment, shows that plastic bags are not only cheaper, but also more environmentally friendly. Their production creates less than half as much atmospheric pollution as the manufacture of paper bags, uses far less energy, and produces 200 times less waste water. In a similar study carried out in the United States to determine whether paper cups were more environmentally friendly than polystyrene disposable cups, plastic again came out ahead.

Although plastic has been shown to be less harmful to the environment than was once thought, the sheer volume of plastic waste poses another problem, and the best method of disposal is not a straightforward choice. More than 85 percent of plastic waste is currently buried in landfill sites or burned. Burning has some advantages because plastic waste contains roughly the same amount of energy as the oil from which it was produced, and this energy can be reused. However, burning must be carefully controlled to avoid giving off dangerous pollutants such as dioxins.

Degradable plastics – which break down into smaller molecules after use – are one possible way to solve the rubbish problem. Others believe recycling plastics is the answer. However, unlike paper and glass, plastic waste is difficult to recycle because it is not uniform – there are many different types of plastic in circulation. Mixtures of plastics tend to be much weaker than individual plastics, and require stabilizers to prevent them from breaking down. But some plastics, such as polyethylene terephthalate (PET), which is used to make bottles and jars, show good potential for recycling into polyester fibre, which could be used as insulation in quilts, pillows and jackets, or as tufting for carpets.

Alternatively, some polymers, such as polyurethane foam, can be hydrolyzed using high-pressure steam to

KEYWORDS

BIODEGRADABLE PLASTIC

CARBONYL GROUP

FLUIDIZED-BED REACTOR

HYDROLYSIS

PLASTIC

POLYETHENE

POLYMER

PYROLYSIS

RECYCLING

■ Plastic bags (made of polyvinyl chloride, PVC) stacked in bales await recycling RIGHT. Another plastic suitable for recycling is polyethylene terephthalate (PET), found in clear soft-drink bottles. BELOW Millions of plastic bottle tops are collected ready to be recycled into plastic sheeting. In addition to common items such as bottles and caps, plastic is also found in larger, less obvious forms, such as in components of cars and airplanes. These "hidden plastics" can be recycled or burned to provide energy.

▽ Mixed plastics are very difficult to recycle. But if they can be sorted either by hand (as shown) or by optical scanners, recycling becomes less expensive. Many communities and cities now have special collection centers where a "first sort" is done.

break them down into their component materials, which can then be reused. Other polymers can be reduced to simple hydrocarbons by means of pyrolysis – "burning" or thermal decomposition in the absence of air. The fact that recycled plastic blends require stabilizers to hold them together gives a clue to the degradability of plastics – however indestructible they may appear. Until recently, chemists were generally more concerned with finding ways to prevent polymers from breaking down, rather than encouraging degradation.

There are two main ways of degrading plastics to help them rot away: biodegradation and photodegradation. Biodegradable plastics (like the polyhydroxybutyrate manufactured under the tradename Biopol) are made out of materials that microbes will digest. Other biodegradable plastics include the biological polymer starch in a matrix of a synthetic polymer like polyethene. When the plastic is discarded, microbes digest the starch and leave a skeleton of polymer that breaks down easily.

Photodegradable plastics contain molecules whose bonds are broken when the material is exposed to sunlight. Some photodegradable polymers incorporate a carbonyl unit (–C=O) into the polymer chain, which absorbs energy from photons of light. When the energy builds up it breaks chemical bonds and fractures the polymer chain.

▷ Pyrolysis, the process of decomposition by heat, can be used to turn discarded mixed plastics back into their constituent molecules. These can then be re-used. Pyrolysis is commonly carried out at temperatures between 400° and 800°C in a fluidized-bed reactor. In this type of reactor, gas is forced upward through a bed of solid particles. This causes the particles to flow in much the same way as a liquid, and improves the efficiency of the energy transfer between the solid and the gas.

Around 50% of the gases produced during pyrolysis are recycled back into the plant to fuel the process. The remaining gases, along with the liquid hydro-carbons produced, can be separated in a fractionation column.

▽ Most of the products of plastics pyrolysis can be used as fuels. They can also be made into monomers to manufacture even more plastics.

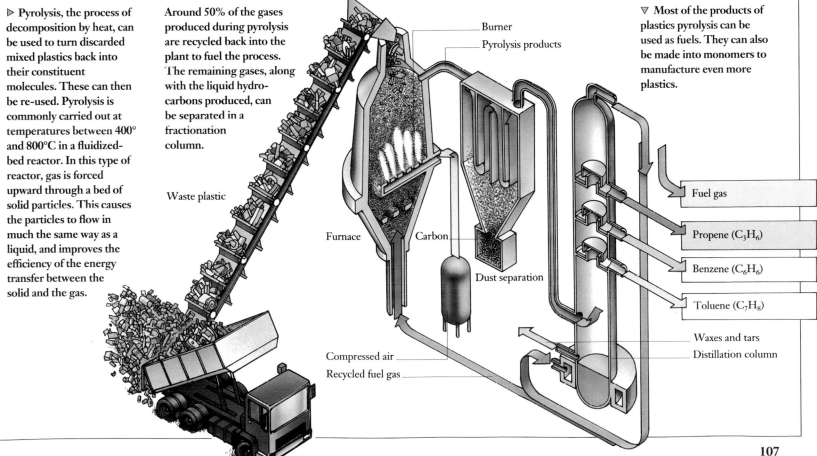

Burner
Pyrolysis products
Waste plastic
Furnace
Carbon
Dust separation
Compressed air
Recycled fuel gas
Fuel gas
Propene (C_3H_6)
Benzene (C_6H_6)
Toluene (C_7H_8)
Waxes and tars
Distillation column

5

CHEMISTRY
of Life

ONLY SIX principal types of molecules form the basis of all
living organisms. These key molecules are made up from
just a handful of chemical elements: hydrogen, oxygen,
carbon, nitrogen, phosphorus and sulfur.

Life on Earth began more than 3.5 billion years ago in the
primitive oceans, which were a dilute soup of ammonia, methanal
(formaldehyde), formic acid, cyanide, methane, hydrogen sulfide and
organic hydrocarbons. These substances in turn condensed from a
primitive atmosphere made up of hydrogen, helium, methane and
ammonia. The first "biochemistry" occurred when energy from the
Sun and sparks from lightning caused the gases to react with each
other to produce more complex molecules with carbon-carbon
bonds. They included several types of amino acids, which are the
building blocks of proteins. Later, through the biochemical reaction
of photosynthesis, primitive plants added oxygen to the atmosphere.
This all-important contribution provided an environment in which
other forms of life, such as animals, could develop.

Biochemistry is still the basis of life. Chemical reactions go on
continuously in all organisms. If they are disturbed by the activities of
microorganisms such as bacteria and viruses, which cause illness,
it may be possible to use chemistry to correct them.

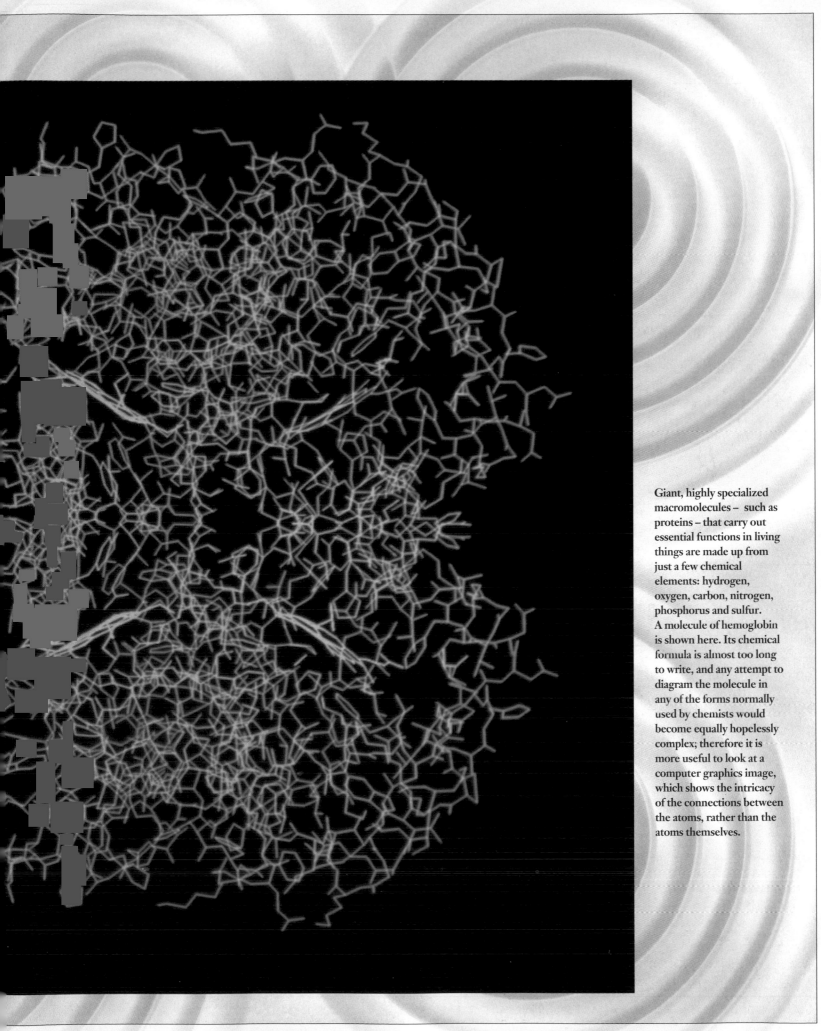

Giant, highly specialized macromolecules – such as proteins – that carry out essential functions in living things are made up from just a few chemical elements: hydrogen, oxygen, carbon, nitrogen, phosphorus and sulfur. A molecule of hemoglobin is shown here. Its chemical formula is almost too long to write, and any attempt to diagram the molecule in any of the forms normally used by chemists would become equally hopelessly complex; therefore it is more useful to look at a computer graphics image, which shows the intricacy of the connections between the atoms, rather than the atoms themselves.

VITAL RAW MATERIALS

OOD provides energy, as well as essential
nutrients – the complex chemical compounds
based mainly on carbon, hydrogen, oxygen and
nitrogen, which all living things need. Like hydrocarbons and other fuels, food contains stored chemical
energy. But whereas energy is released from
hydrocarbons very quickly during combustion, or
burning, energy is released in a controlled way from
food during a different type of oxidation process
known as respiration.

Respiration is a
sequence of complex
chemical reactions by
which food is broken down
by reaction with oxygen.
As with combustion, the
products of respiration are
energy, carbon dioxide and
water. Some of the energy
is released as heat, and
some is stored in adenosine
triphosphate (ATP)
molecules, which are broken down into adenosine
diphosphate (ADP) to release stored energy as needed.
The chemical reactions in respiration can be speeded
up or slowed down to meet different energy demands.

Before respiration can take place, food must be
broken down. Food is physically broken down by
chewing and by churning in the stomach, and then
chemically broken up in the digestive tract by enzymes
– proteins that act as catalysts in living systems.

In order to stay healthy, people must eat seven
different types of compounds. These essential
nutrients include vitamins and mineral salts; fiber;
water; lipids, in the form of fats and oils; carbohydrates, in the form of starches and sugars; and
proteins. Vitamins and minerals protect the body from
disease and are needed in order to make use of other
nutrients. Minerals also provide important elements
needed to make more complicated molecules. Fiber
helps to dispose of waste material. Water, which
makes up roughly 75 percent of the human body,
provides the essential solvent in which most
biochemical processes take place.

Lipids (fats and oils) – which are made up only of
carbon, hydrogen and oxygen – are important sources
of stored energy. Lipids consist of a stem made up
of glycerol with three fatty acid molecules attached.
Each of the fatty acids is made up of a long carbon
chain which has mainly hydrogen atoms attached to it.
The glycerol and fatty acids are released during
digestion and recombine as new fats which can be
stored in various body tissues.

KEYWORDS

AMINO ACID
CARBOHYDRATE
ENZYME
FAT
GLUCOSE
MINERAL
OXIDATION
PROTEIN
VITAMIN

■ Animals obtain energy
from food via a complicated
sequence of oxidation
reactions known as
respiration. During
respiration, food – usually
in the form of glucose – is
broken down in an animal's
body BELOW. Enzymes –
proteins that act as catalysts
in living systems – play an
important role in glycolysis,
the first stage in this
process. RIGHT An ATP
molecule (pink) is shown
bound to a glycolytic
enzyme. The green ribbon
is a carbon chain; oxygen,
carbon and nitrogen
atoms of ATP appear
in red, white and
blue respectively.

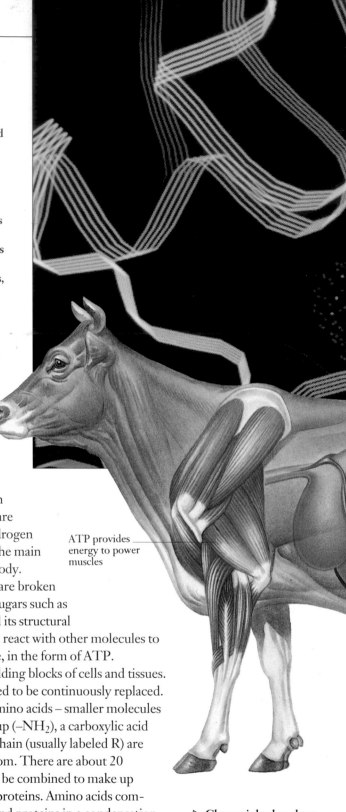

ATP provides
energy to power
muscles

Carbohydrates, such
as starches and sugars, are
made up of carbon, hydrogen
and oxygen. They are the main
energy source for the body.
These large molecules are broken
down to form smaller sugars such as
glucose ($C_6H_{12}O_6$) and its structural
isomer, fructose, which react with other molecules to
provide an energy store, in the form of ATP.

Proteins are the building blocks of cells and tissues.
As a result, proteins need to be continuously replaced.
They are made up of amino acids – smaller molecules
in which an amino group ($-NH_2$), a carboxylic acid
($-COOH$), and a side chain (usually labeled R) are
attached to a carbon atom. There are about 20
amino acids, which can be combined to make up
thousands of different proteins. Amino acids combine to form peptides and proteins in a condensation
reaction during which the carboxylic acid group on
one amino acid joins on to the amino group on the
next, and a water molecule is lost. Proteins can contain
more than 4000 amino acid units.

It is not necessary to eat proteins identical to the
ones that are being replaced. Proteins are broken
down during digestion into their basic amino acids,
which can join together to make different proteins.
One amino acid can be converted into another, and
some can be made from carbohydrates. But there are
eight essential amino acids that cannot be manufactured by the body, and these must be taken in as food.

▷ Glucose is broken down
via a series of reactions
during respiration that
produce molecules of ATP
(adenosine triphosphate),
the energy currency of
living cells. The first step in
respiration is glycolysis.
During this process,
enzymes catalyze the
breakdown of glucose into
molecules of pyruvic acid.
The pyruvic acid is then
involved in a complex cycle
of enzyme-controlled

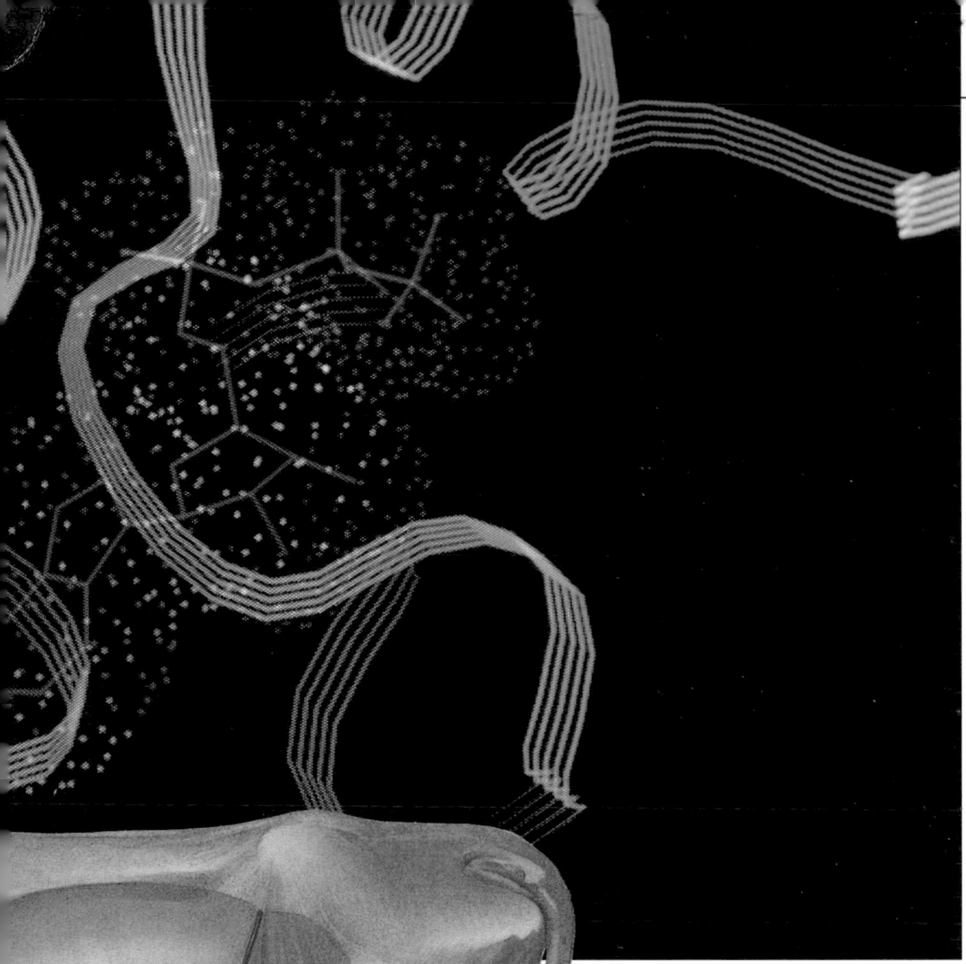

▽ **ATP**, the main energy-carrying molecule, has a core surrounded by triphosphate groups. The core of **AMP** (adenosine monophosphate) is based on adenine molecules, composed of two carbon-nitrogen rings attached to molecules of the five-carbon sugar ribose. The triphosphate groups are attached to the core by high-energy bonds. When one of these is broken to produce ADP, around 7 kcal of energy is released per mole of ATP.

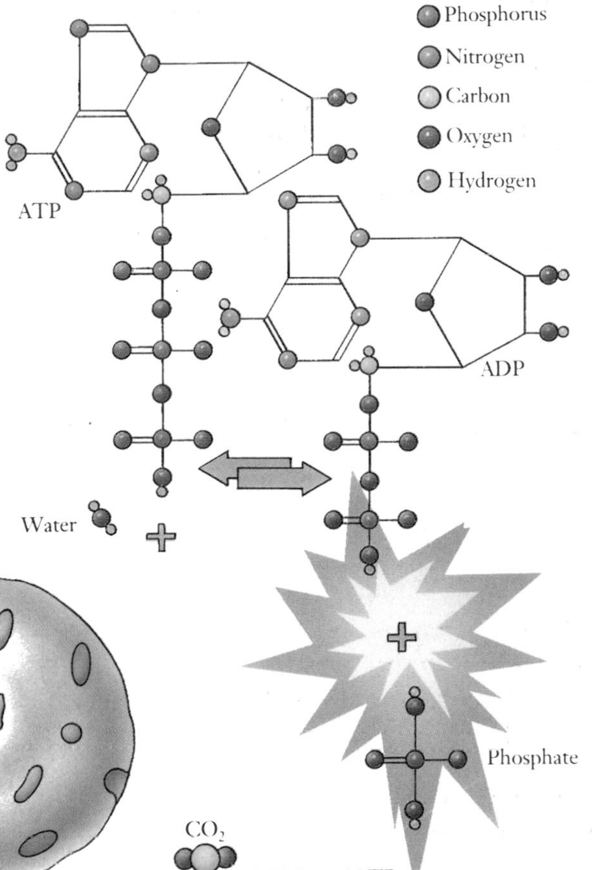

- Phosphorus
- Nitrogen
- Carbon
- Oxygen
- Hydrogen

ATP

ADP

Water

Phosphate

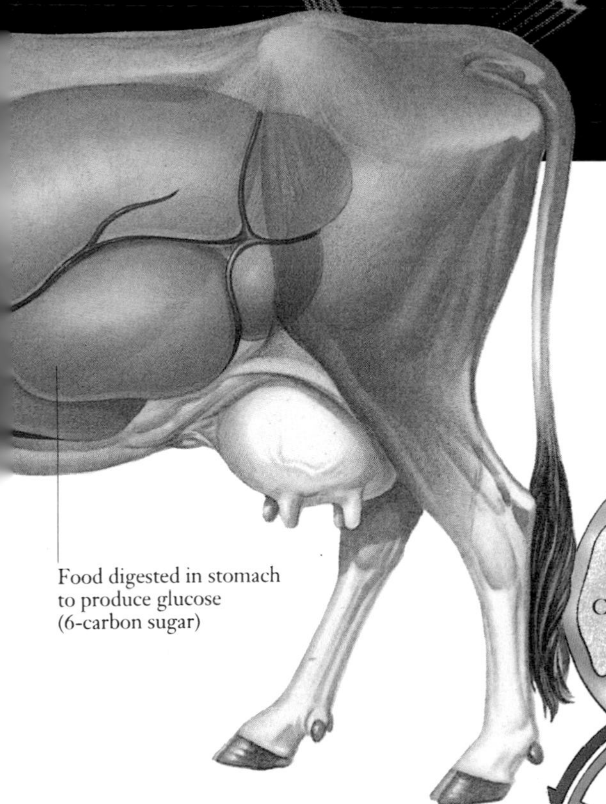

Food digested in stomach to produce glucose (6-carbon sugar)

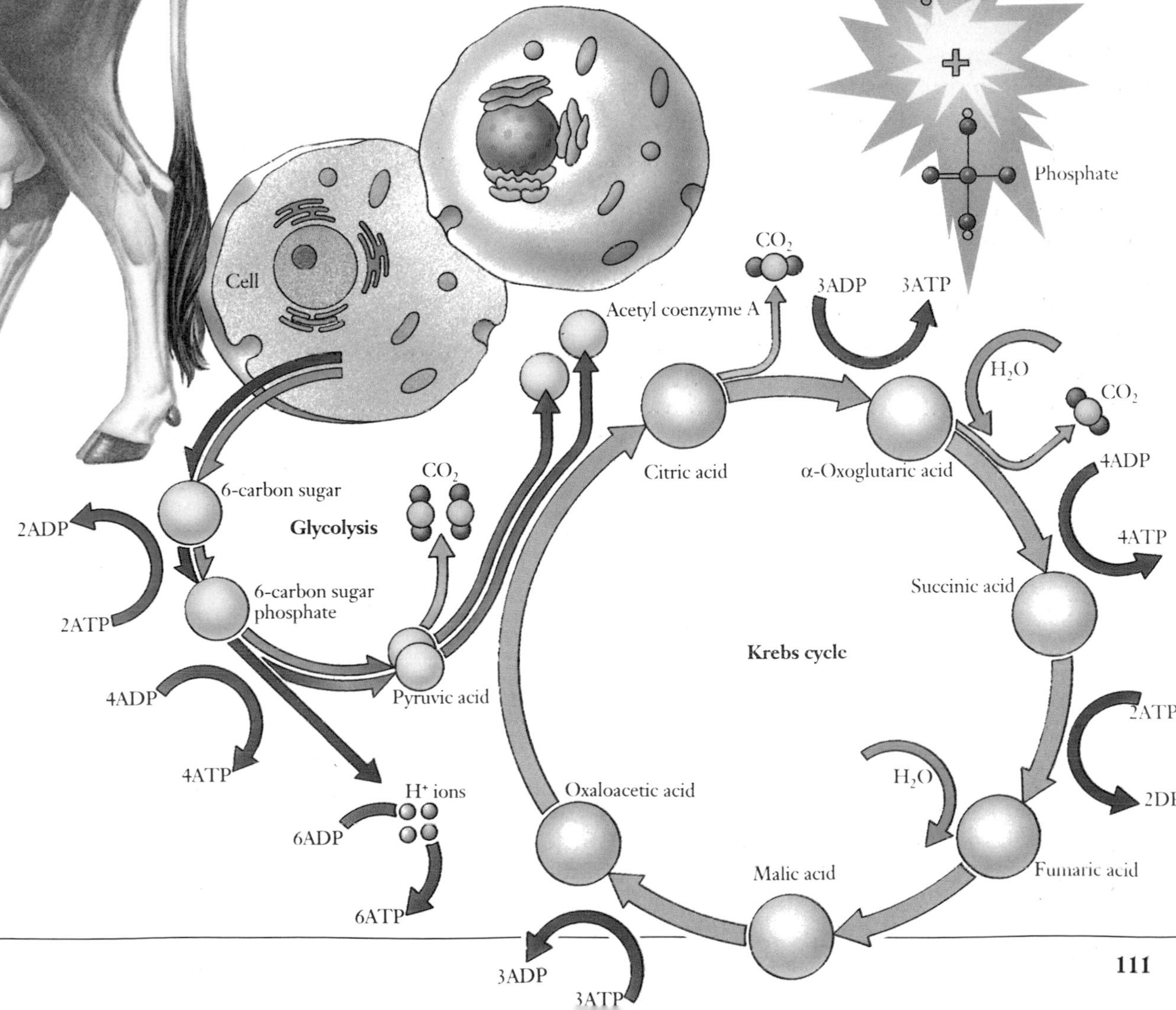

Cell

reactions known as the Krebs, or citric acid, cycle. During the Krebs cycle, some amino acids are synthesized, along with molecules of the hydrogen ion carrier NAD. These are involved in a further series of reactions to generate molecules of ATP. Energy is released by ATP when it is converted to ADP (adenosine diphosphate) by the removal of a phosphate group.

6-carbon sugar
2ADP
Glycolysis
2ATP
6-carbon sugar phosphate
4ADP
4ATP
CO_2
Pyruvic acid
H^+ ions
6ADP
6ATP

Acetyl coenzyme A
CO_2
3ADP 3ATP
Citric acid
α-Oxoglutaric acid
H_2O
CO_2
4ADP
4ATP
Succinic acid
Krebs cycle
2ATP
H_2O
2DP
Oxaloacetic acid
Malic acid
Fumaric acid
3ADP
3ATP

111

LIVING CHEMISTRY

Aʟʟ living things are chemical machines. Chemical reactions within the body provide organisms with energy, get rid of waste products, and keep them healthy. Almost all biological processes rely on catalysts in the form of proteins known as enzymes.

Enzymes are highly specialized proteins that regulate metabolism – all the chemical processes that occur in a cell or organism. They act as organic catalysts that speed up chemical reactions in living cells. Without their catalyzing power, many biochemical reactions would proceed too slowly to sustain life.

The action of enzymes is very specific. In fact, most enzymes are involved in just one chemical reaction. Their structure explains why. Like all proteins, they are composed of folded chains of amino acids held together by weak dipole-dipole and hydrogen bonds. The surface of an enzyme contains specially shaped depressions called active sites in which catalysis occurs.

An enzyme becomes inactive if its shape is changed. Enzymes are inactivated, or denatured, by changes in temperature, which cause the molecules to vibrate more vigorously and thus break intermolecular bonds; or by changes in pH, which interfere with the ionic interactions that hold the folded chain in shape. There are thousands of different enzymes. Most work during metabolism as part of a chain reaction, or catalytic cycle. In general, the product of one enzyme-induced reaction becomes the starting point for the next.

Hormones are the chemical messengers used by the central nervous system in animals to control and regulate body chemistry. In plants, phytohormones do a similar job. Hormones are produced in one part of the body, then transported – usually via the bloodstream – to a target cell in another part where they act by regulating preexisting processes.

There are three main types of hormones. Peptide, polypeptide and protein hormones such as somatotropin, the growth hormone, are made up of chains of amino acids of various lengths. Hormones like epinephrine (adrenaline), which prepares the body for fast action in an emergency, are based on amino acid derivatives known as amides, which are organic compounds in which one or more of the hydrogen

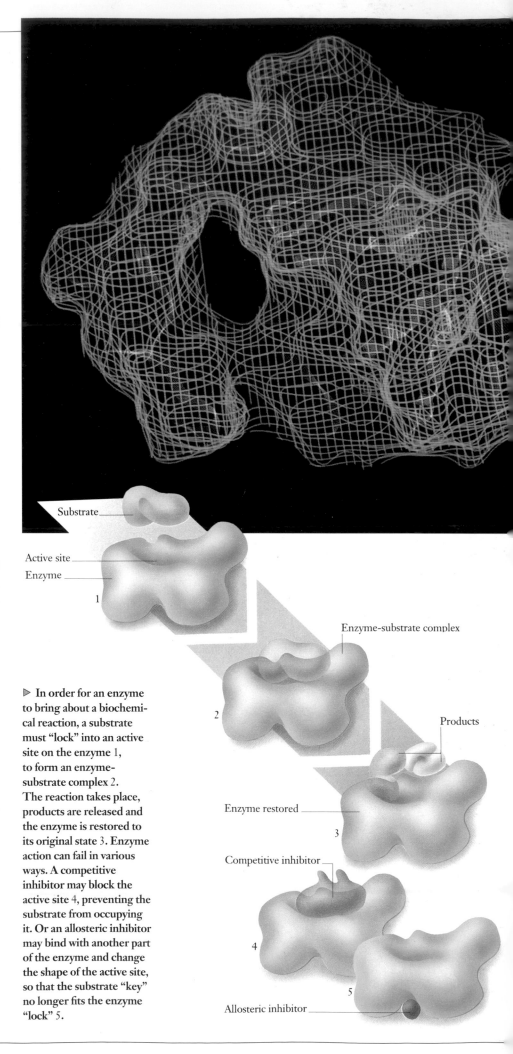

Substrate

Active site

Enzyme

1

Enzyme-substrate complex

2

Products

Enzyme restored

3

Competitive inhibitor

4

5

Allosteric inhibitor

▷ In order for an enzyme to bring about a biochemical reaction, a substrate must "lock" into an active site on the enzyme 1, to form an enzyme-substrate complex 2. The reaction takes place, products are released and the enzyme is restored to its original state 3. Enzyme action can fail in various ways. A competitive inhibitor may block the active site 4, preventing the substrate from occupying it. Or an allosteric inhibitor may bind with another part of the enzyme and change the shape of the active site, so that the substrate "key" no longer fits the enzyme "lock" 5.

◁ An enzyme molecule consists of much-folded chains of amino acids that make up the polypeptides of the enzyme's protein structure. The resulting geometry creates an active site, into which a substrate fits like a key fitting a lock. This close contact usually puts the chemical bond to the substrate under stress, making the bond break and splitting the substrate into the two desired products. The enzyme shown here is an amylase, a category of enzymes that break down starch, glycogen and other polysaccharides.

▽ The two chief types of hormone are either steroid-based (such as sex hormones) or protein-based (like insulin). Steroids are soluble in fats and can therefore pass directly through the target cell membrane, to be picked up by a receptor and carried to the nucleus. This is indicated by the green arrows. Proteins cannot pass through the cell membrane. Instead they attach to a receptor in the membrane itself before being carried into the cell, as shown by the purple arrows.

atoms in ammonia (NH_3) have been replaced by an acyl group (-RCO). Steroid hormones, such as the sex hormones, are based on the linked benzene rings that make up the molecule cholesterol.

Hormones perform a crucial role. For instance, diabetics do not produce enough of the hormone insulin, and so are unable to take in glucose from their blood. Diabetes has many serious symptoms and can lead to death if not treated.

The polypeptide hormones insulin and glucagon, and the amide-based hormone epinephrine, work together to control glucose levels in the blood. Glucose is stored in the body as the compound glycogen. When blood glucose levels fall below normal, glucagon and epinephrine are released into the bloodstream. They act in the liver to increase the rate at which glycogen is converted to glucose. This leads to an increase in blood glucose levels.

If blood glucose levels become too high, insulin is released. It stimulates the liver and muscles to convert blood glucose into glycogen, so that blood glucose levels fall. In this feedback loop, the amounts of insulin and glucagon produced are determined by the level of glucose in the blood, while the overall level of glucose is determined by the balance between these two hormones.

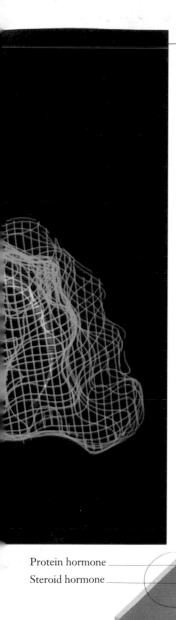

Protein hormone
Steroid hormone
Blood vessel
Bloodstream
Target cells
Receptor
Hormone
Membrane
Nucleus
Target cell
Membrane
Hormone
Receptor
Target cell

△ Sex hormones bring about profound changes in developing young animals. Both sexes of young gorillas, for example, differ externally only in their genitals. But as they enter puberty, hormones from the sex glands develop secondary sexual characteristics that distinguish adult males from adult females. The males grow silver fur on their backs, and the females develop mammary glands to produce milk for offspring.

MEDICAL DRUGS

DISEASES result from chemical changes that disrupt the life processes of organisms. Chemotherapy – the use of chemicals to fight disease – is a powerful tool to bring the chemical machine back into balance. Drugs play an important part in this chemical warfare. All drugs work by altering the biochemical processes in either the disease-causing organism or the organism affected by the disease.

Different types of drugs act in different ways to fight disease. Vaccines prevent illness by stimulating the body's immune system to develop special proteins called antibodies that attack disease-causing organisms. Other medicines affect the biochemical pathways in the attacking organism. This can involve, for example, blocking the action of enzymes and thus preventing biochemical reactions from going out of control; or preventing hormones from delivering their chemical messages, and thus blocking the response of a cell to a hormone. Antibiotics and sulfa drugs, both hailed as "wonder drugs" in their time, take this approach.

Antibiotics are extracted from living micro-organisms and selectively destroy disease-causing bacteria, often by inhibiting the action of important enzymes in the bacteria. Penicillin stops the growth of new bacteria by inhibiting the action of an enzyme responsible for constructing the bacteria's cell wall. Sulfa drugs, in contrast, are produced in a test tube and inhibit bacteria from synthesizing folic acid, an essential nutrient. This prevents the bacteria from reproducing and gives the body's defense mechanisms a better chance of killing off the invaders. Sulfa drugs are harmless to humans because mammals cannot synthesize folic acid, and must include it in their diet.

Like enzymes, drugs depend on a "lock and key" mechanism to work. The drug molecule must fit exactly into a receptor on the molecule whose chemistry it hopes to influence. It is also critical that the drug is delivered efficiently to the part of the body it is supposed to affect. For example, a pill must be designed so that it is released and absorbed in one part of the digestive system rather than another. In some cases it is possible to attach a drug molecule to an antibody, allowing the drug to target diseased cells directly. This approach is used to treat cancerous tumors without harming surrounding cells.

KEYWORDS

ALCOHOL
BARBITURATE
BIOCHEMISTRY
CHEMOTHERAPY
DRUG
ENZYME
HORMONE
MOLECULAR MODELING
SYNTHETIC DRUG

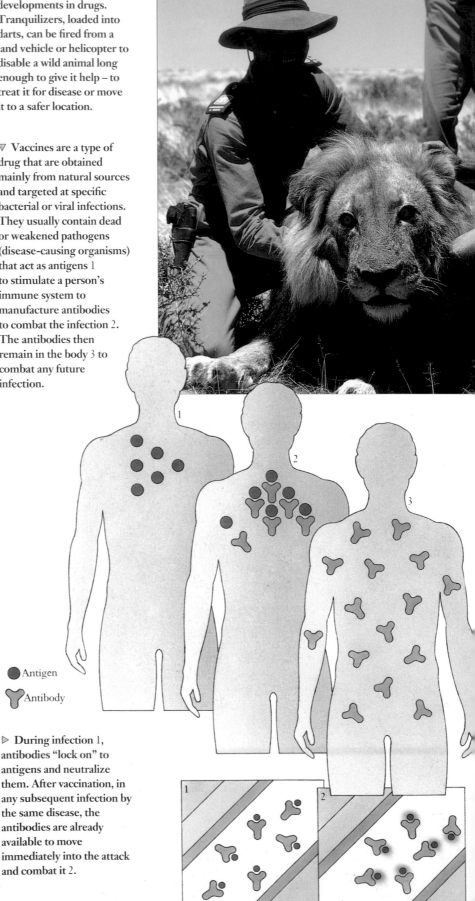

▷ Veterinary medicine has benefited from recent developments in drugs. Tranquilizers, loaded into darts, can be fired from a land vehicle or helicopter to disable a wild animal long enough to give it help – to treat it for disease or move it to a safer location.

▽ Vaccines are a type of drug that are obtained mainly from natural sources and targeted at specific bacterial or viral infections. They usually contain dead or weakened pathogens (disease-causing organisms) that act as antigens 1 to stimulate a person's immune system to manufacture antibodies to combat the infection 2. The antibodies then remain in the body 3 to combat any future infection.

● Antigen

Y Antibody

▷ During infection 1, antibodies "lock on" to antigens and neutralize them. After vaccination, in any subsequent infection by the same disease, the antibodies are already available to move immediately into the attack and combat it 2.

▽ The beta-blocker drug propanolol acts against the natural substance norepinephrine (noradrenalin), which stimulates heartbeat and raises blood pressure, by blocking receptor sites in the heart where it can bind and take effect.

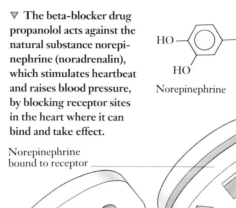

Norepinephrine

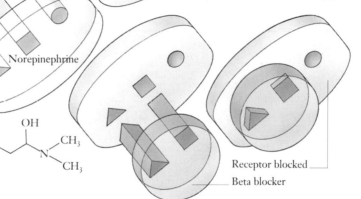

Norepinephrine bound to receptor

Receptor

Norepinephrine

Propranolol

Receptor blocked

Beta blocker

Not all drugs are medicines – some can unbalance body chemistry and act as poisons. Drugs like alcohol and cocaine depress the central nervous system by interfering with the activity of neurotransmitters and receptors on the nerve cells. These drugs can relieve pain and tension, but they also slow reaction times and impair judgment. When taken in large amounts, or in combination with other drugs, the effects can be fatal. Chronic overuse can cause physical deterioration, such as cirrhosis of the liver in alcoholics. Widespread misuse of barbiturates, another class of drugs that reduce the activity of the central nervous system, has made it necessary to restrict their use. In some countries they are now routinely prescribed only for epilepsy.

▽ Many drugs in tablet form are dispensed in bubble packets to prevent them from deteriorating, or to make them available in "sets" to suit a prescribed dosage – for example, as a certain number of tablets per week or month.

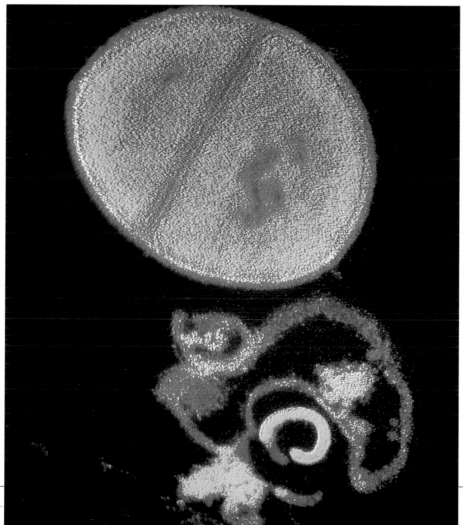

◁ There is now a wide range of antibiotics to combat an equally wide range of bacterial infections. Some – like the original antibiotic drug penicillin – are derived from natural sources (often molds and other microorganisms). Other modern antibiotics are synthetic; they are laboratory copies of natural substances or tailor-made to combat a particular infection. As shown here, they work by breaking down the cell walls of the bacteria.

Natural Drugs

By understanding the effects of diseases at the molecular level, chemists are able to search for – or to design and manufacture – specific substances to fight specific diseases. Some of the most widely used drugs were developed by extracting, refining and purifying active ingredients derived from plants. Aspirin, which is based on 2-hydroxybenzoic (or salicylic) acid – a compound found in the bark of the willow tree – is now produced by introducing an extra functional group into the phenol molecule to produce 2-ethanoyl-hydroxybenzoic acid. Digitalis is still made by refining an active compound found in a species of foxglove.

KEYWORDS

ALKALOID

CHIRALITY

DRUG

ISOMERISM

PHARMACOLOGY

RACEMIC MIXTURE

Random screening of alkaloids (nitrogen-containing organic compounds found in some plants) has resulted in the identification of many useful active compounds. The search for new compounds is concentrated in areas such as tropical rainforests, where there is a great diversity of plants whose properties have not yet been explored.

This approach led to the development of the drug captopril, which is widely used to treat high blood pressure. The inspiration for its development was the discovery that certain proteins in the venom of the Brazilian arrowhead viper act as an inhibitor to block the action of an enzyme which is a key factor in raising blood pressure. By using the structure of the proteins in the venom as a starting point, chemists were able to synthesize a successful drug to lower blood pressure.

As well as the composition of a drug, chemists also have to consider its shape and orientation. Many biological molecules are chiral – they exist in right- and left-handed forms, or isomers. Although they often have the same physical properties, and seem to behave chemically in the same way in the test tube, they can have very different effects in the body, where they react with other chiral molecules. For example, right-handed amino acids react with the taste buds to taste sweet, whereas left-handed amino acids often taste bitter, or are inactive and have no taste at all.

Because it is expensive to separate the different isomers, many medicines are sold as racemic mixtures, in which there are equal numbers of left- and right-handed forms. Careful testing is necessary to make sure that they are not harmful. In the case of the drug thalidomide, the active isomer was a mild sedative, but when taken by pregnant women, the inactive isomer caused severe damage to their unborn children.

□ Plants still provide a more economical source of modern drugs than synthetic chemicals. Examples include the heart drug digitalis, which is still obtained from foxgloves 1; caffeine from the coffee plant 2; morphine from the opium poppy 3; and atropine from the deadly nightshade 4. Plants are also promising sources for new drugs. Pharmaceutical companies pay large sums for the rights to search rainforests for new plants which may contain the active compounds for the drugs of the future.

1 Foxglove (digitalis)

2 Coffee (caffeine)

3 Opium poppy (morphine)

4 Deadly nightshade (atropine)

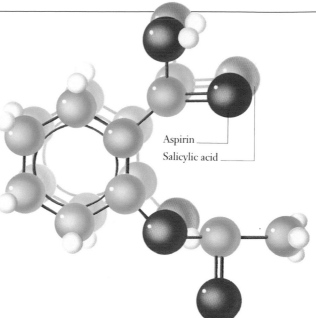

◁ Salicylic acid, extracted from the bark of the willow tree, was used in the past to relieve fevers and pain. The modern drug aspirin consists of an ethanoyl (acetyl) derivative of salicylic acid. Its sodium salt is also a mild painkiller and is used to treat rheumatism.

Aspirin
Salicylic acid

● Carbon
● Hydrogen
● Oxygen

△ Many new drugs are now invented and tested on a computer screen with the aid of powerful graphics programs. These new compounds, like all other drugs, must undergo extensive clinical trials before they are approved.

◁ A market stall in Chile offers a wide range of natural medicinal products. Drug companies are just as interested in the goods as the local people. Local markets often prove to be a rich source of new drugs.

At the start of every synthesis of a new drug is a lead compound – something that shows some promise of giving the desired result. Today, with the help of powerful computer graphics programs, chemists can use their knowledge of the chemical basis of disease and the activity of lead compounds to build up molecules and test their possible effects on the computer screen. They can, in effect, design drugs by synthesizing them on screen.

In spite of the increasing success of drug design based on the use of computers, and an understanding of the molecular basis of drug molecules and disease, finding, developing and marketing a new drug is still a very expensive and risky business. Only about 1 in 10,000 of the compounds synthesized survive the rigorous testing procedures carried out before a drug can be made commercially available.

TESTING FOR DRUGS

SIMPLE medical testing kits bring biochemistry out of the laboratory and make it possible for people to determine the concentration of various substances in their own bodies very quickly and relatively accurately without going to a laboratory. Other portable tests make it possible for non-medical professionals, such as law enforcement personnel, to obtain information about drugs and alcohol in a person's system, without waiting for the result of a more expensive laboratory test. Behind their seeming simplicity, the kits conceal some clever chemistry.

Medical testing kits have many applications, including self-monitoring for diabetics and on-the-spot analysis of a motorist suspected of drunk driving. The breathalyzer test to determine the level of ethanol (ethyl alcohol) in the blood, or blood alcohol content (BAC), relies on a single chemical reaction: the reduction of potassium dichromate (VI) to chromium (III) by ethanol. The orange crystals of potassium dichromate are reduced and turn green when they oxidize ethanol to ethanal (acetaldehyde) and ethanoic (acetic) acid.

Because the oxidation of ethanol involves electron transfer, the concentration of ethanol can also be measured by the voltage of an electrolytic cell that incorporates the reaction. In one type of breathalyzer, phosphoric acid is held in a porous plastic material between two electrodes. Oxygen is reduced to water at one electrode, and ethanol is oxidized to ethanoic acid at the other.

▷ In the United Kingdom, around 10 percent of all horses racing are tested for drugs, some selected at random and some chosen because they performed either better or worse than expected. Urine samples are preferred to blood samples because the concentrations of drugs and their metabolites are generally higher than in blood. Tests used to detect prohibited substances include gas chromatography, high-performance liquid chromatography, mass spectrometry and enzyme-linked immunoassays. Of the hundreds of prohibited substances, the most commonly found include phenylbutazone, a non-steroidal anti-inflammatory drug which masks stiffness or joint injury; caffeine, which acts as a stimulant; and acepromazine, a sedative which can be used to slow down a racehorse or to calm a horse in a dressage event.

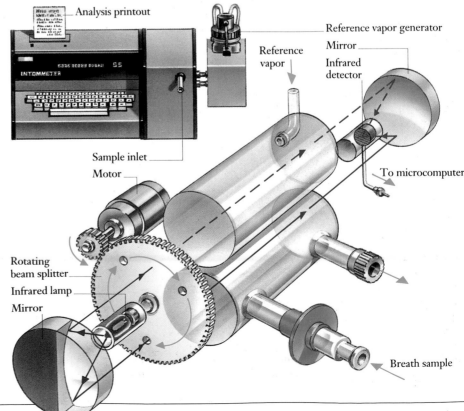

Analysis printout

Reference vapor generator

Mirror

Reference vapor

Infrared detector

Sample inlet

Motor

To microcomputer

Rotating beam splitter

Infrared lamp

Mirror

Breath sample

However, these roadside breathalyzer tests are not accurate or reliable enough to be used as evidence in court. To obtain a more accurate determination of ethanol levels, blood or urine samples are analyzed by gas-liquid chromatography. Ethanol (alcohol) concentration can also be accurately measured using an instrument that analyzes the ethanol in the breath by absorption of infrared radiation.

People with diabetes must keep a careful watch on their blood glucose levels, and many find that simple reagent strips to test for glucose in blood or urine are a valuable aid. The strips are impregnated with four reagents. The first is glucose oxidase, an enzyme that catalyses the reaction between glucose and oxygen to produce gluconic acid and hydrogen peroxide. There is also an indicator, such as orthotoluidene, which is colorless in its reduced form, but turns into a colored form when oxidized. The strips also contain peroxidase, an enzyme that catalyses the oxidation of the indicator by hydrogen peroxide; and a buffer, a mixture of chemicals that keep the reagents at a fixed pH during the test.

To carry out the test, the strip is dipped in urine, or blood is dropped onto it. If glucose is present, the strip changes color, and the level of glucose can be estimated by comparison with a special color chart. Small reflectometers to measure the color change are also available to give more accurate readings.

Electrochemistry is another way of obtaining accurate readings of the glucose levels revealed by enzyme reactions. One type of glucose monitor also uses the enzyme glucose oxidase (as in the test strips). This enzyme reacts with one of the three common forms of glucose, β (GK)–D–glucose, and the reaction causes a change in potential difference in a cell, which can be measured using an electrochemical sensor.

Another brand of test strip uses electrodes, which contain an enzyme complex with glucose oxidase and ferrocene. When a drop of blood is placed on the sensor, glucose is oxidized to gluconolactone and glucose oxidase is reduced. The electrons released are absorbed by the mediator ferrocinium, to form ferrocene. The ferrocene is then oxidized back to ferrocinium at the electrode. The current of electrons which results from this redox reaction is measured by the electrode and is proportional to the glucose level in the blood.

◁ The roadside breath-alyzer test used by police takes advantage of the fact that ethanol (alcohol) is volatile enough to pass through the blood into the air in the lungs. A suitable sample is provided by "blowing into the bag". Alcohol levels are then determined using a simple oxidation-reduction reaction, which causes a color change in crystals of potassium dichromate. Alternatively, a suspect may be asked to exhale into a machine in which the absorption of infrared radiation is used to measure the amount of alcohol present. The machine compares the sample with the known alcohol content of a reference vapor.

△ Because the formation of urine tends to concentrate traces of drugs, athletes in major sporting events such as the Olympics are routinely subjected to urine testing to detect drug use.

Sensitive methods such as thin-layer or gas chromatography are used for finding even small traces of drugs in the urine.

6

CHEMISTRY
and Color

DYES AND PIGMENTS are only two of the products of
the colorful science of chemistry. Color is not only
produced by chemists but is also used to reveal chemical
reactions, as in an acid-base indicator.

All objects appear colored because they absorb certain wavelengths of light, and reflect or emit others which are visible.
Fluorescent molecules absorb ultraviolet or invisible radiation and
re-emit visible light. In organic compounds, color is due to
chromophores – sequences of atoms linked by double bonds in either
chains or rings, which absorb specific wavelengths of light, and
reflect the rest. By altering the number or sequence of the double
bonds, chemists can vary the colors of organic compounds.

At the basis of color is light, which can provide the activation
energy needed to begin many chemical reactions. Study of the
emission and absorption of different frequencies of radiation provides
the basis for the spectroscopic techniques used for chemical analysis.

Photochemistry is the study of reactions caused by light. Photochemical reactions make photography possible, cause our skin to
darken and pollution smog to form. They also form the basis of life
on Earth: plants capture energy from the Sun in a series of
photochemical reactions which take place during photosynthesis.

The chemist's palette now has thousands of colors, due to an improved understanding of the structure of chromophores – molecules that absorb specific wavelengths of light. White light contains all the wavelengths of visible light and therefore the whole range of colors. When wavelengths corresponding to a given color are absorbed, and thereby removed from the white light, the complementary color is seen. This is why an object that absorbs in the red wavelength region appears green, and one that absorbs in the blue region appears orange.

PAINTS, PIGMENTS AND INKS

THE BLUE of the sky is due to the scattering and bending of light – but the blue color of a wall or the color of a car is due to the presence of colored compounds known as pigments. These interact with light to emit different frequencies of radiation which are perceived as color. Pigments are insoluble substances that can be spread as surface layer, or mixed in with the bulk of a material, to give color. Some pigments, such as chlorophyll, the green pigment in plants, are organic compounds, but the majority of the ones used in paints and inks are inorganic. Their colors are often due to the chemical properties of the transition metals, which are important components of many pigments. In contrast to metals like sodium and magnesium – which always have the same number of electrons available for bonding in their outer shell, so have only one oxidation state – each transition metal can exist in two or more oxidation states. Many compounds involving transition metals appear in a wide variety of colors. The color depends on the oxidation state of the metal ion and on the type and arrangement of the other molecules, or ligands, which bond to it.

Transition metals are also used in fluorescent and phosphorescent paints. Fluorescent paints contain zinc and cadmium sulfides along with organic dyes. They absorb ultraviolet light and re-emit it as visible light when irradiated. Phosphorescent paints incorporate phosphors, such as zinc, copper or strontium sulfate, which continue to glow after irradiation has ceased.

Pigments can be applied to surfaces using paints or to paper using ink. Both add color to surfaces in a similar way. Paints have two basic components: the vehicle and the pigment. The vehicle, or binder, is dissolved in an extender to form the liquid part of the paint, which polymerizes to provide the bonding and protective film. The vehicle controls the flow properties of the coating and helps to improve its hardness and toughness. Oil-based paints use natural polyunsaturated oils such as fish oil or linseed oil as a vehicle. They also contain a solvent, to dissolve the oil or resin. Synthetic alkyd resins are generally used to replace the natural oil. The resins are produced by a condensation reaction between certain alcohols and acids, and composed of polymers made up of a long chains containing carbon, hydrogen and oxygen atoms with unsaturated carbon-hydrogen side chains.

Water-based paints, also known as latex or emulsion paints, contain highly polymerized resins such as polyvinyl acetate (PVA) or a copolymer such as a styrene-butadiene resin, formulated as an emulsion in water. For use outdoors, latex paints generally contain a high proportion of resin to provide a film that is stable to weathering.

Paints dry when the solvent they contain – whether water or an organic solvent – evaporates. At the same time, the polymers in the paint start to oxidize to form a film. This happens when the alkyd resins in paint come into contact with air and the double bonds in the side chains react with oxygen in the air and form cross-links between unsaturated sites on neighboring polymers. The cross-links between the polymer chains are what hold the paint molecules together in the film. This type of reaction is also important in adhesives, which are composed of many of the same types of molecules that are used in paint.

INKS AND COLOR

Inks are used to apply pigments to paper. In the past, they were available only in black and were based on a suspension of carbon black (a kind of soot) or on a mixture of an iron salt and tannic or gallic acid. Modern inks contain colorfast pigments to prevent fading and are available in a wide range of colors, such as red (iron phthalocyanine, **1**) and violet (dioxazine, **2**).

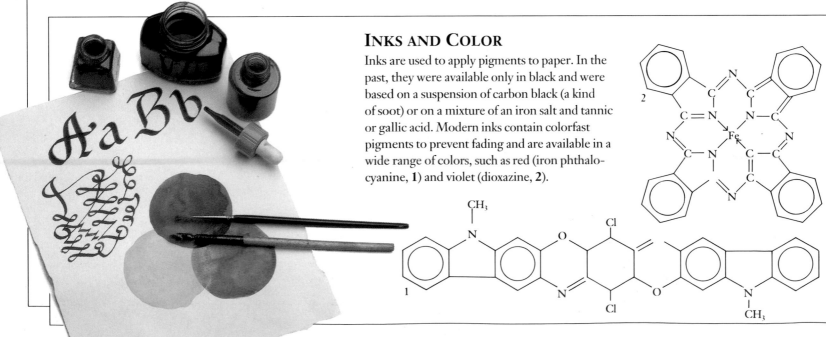

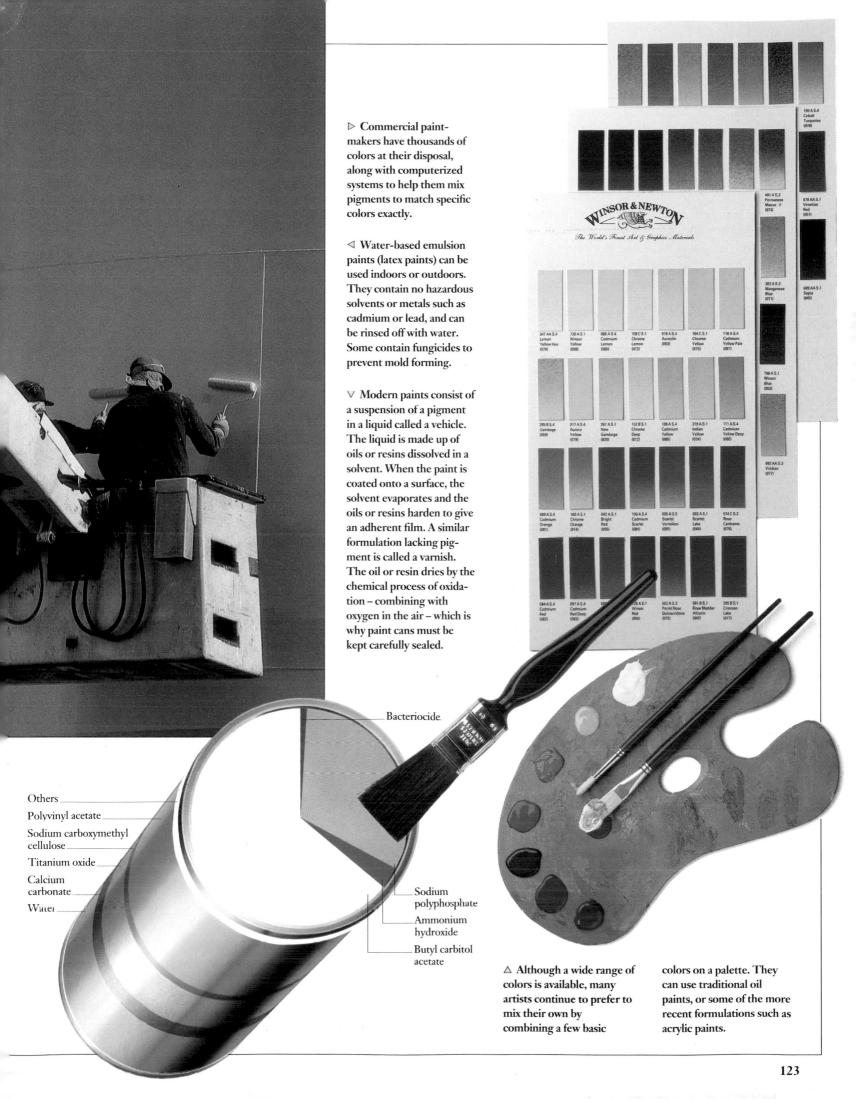

▷ Commercial paint-makers have thousands of colors at their disposal, along with computerized systems to help them mix pigments to match specific colors exactly.

◁ Water-based emulsion paints (latex paints) can be used indoors or outdoors. They contain no hazardous solvents or metals such as cadmium or lead, and can be rinsed off with water. Some contain fungicides to prevent mold forming.

▽ Modern paints consist of a suspension of a pigment in a liquid called a vehicle. The liquid is made up of oils or resins dissolved in a solvent. When the paint is coated onto a surface, the solvent evaporates and the oils or resins harden to give an adherent film. A similar formulation lacking pigment is called a varnish. The oil or resin dries by the chemical process of oxidation – combining with oxygen in the air – which is why paint cans must be kept carefully sealed.

Bacteriocide

Others

Polyvinyl acetate

Sodium carboxymethyl cellulose

Titanium oxide

Calcium carbonate

Water

Sodium polyphosphate

Ammonium hydroxide

Butyl carbitol acetate

△ Although a wide range of colors is available, many artists continue to prefer to mix their own by combining a few basic colors on a palette. They can use traditional oil paints, or some of the more recent formulations such as acrylic paints.

DYES AND DYEING

PIGMENTS and dyes both give color, but in pigments the color is on the surface. Pigments stick to surfaces to give them color, whereas dyes attach themselves chemically to the molecules that they color. Sometimes ionic or covalent bonding is involved, but often the attachment is the result of hydrogen bonding, or weaker intermolecular forces.

In contrast to pigments, most dyes are soluble, and most are aromatic organic compounds. Many naturally occurring organic dyes are derived from plants and animals. The red dye cochineal is extracted from a species of insect. A brilliant orange dye is made from the dried stigmas (pollen-collecting organs) of the saffron crocus. The blue dye indigo is derived from compounds known as leucoanthycyanidins, which occur in plants of the genus *Indigofera*. The red dye alizarin comes from the root of the madder plant.

Many natural dyes stick fast to cloth only with the aid of a mordant, a metal compound which attaches itself to the cloth under alkaline conditions, and then binds to the dye molecules. By using different metals in the mordant it is often possible to vary the color of the dyed cloth. For example, when alizarin is used with a mordant containing tin(II), the result is a pink color. With iron(III), the cloth is dyed brown.

Synthetic dyes can be designed to give a wider range of colors and properties than are available in natural dyes. Synthetic dyes are generally based on aromatic hydrocarbons such as benzene, toluene and naphthalene, and their derivatives such as aniline. Some of the first synthetic dyes to be developed were the azo dyes. They are still used to produce a wide range of mainly yellow, orange or red shades.

Azo dyes are very stable compounds which do not fade or loose their color. Many contain sulfonic acid groups ($-SO_3-$) in their structure to make then soluble in water and to help the dyes bind tightly to the large complex molecules in textile fibers.

The color in a dye molecule is due to a group of atoms known as a chromophore. In azo dyes the chromophore is made up of aromatic (benzene) rings linked by double N=N bonds. The chromophores are often part of an extended arene complex, a system in which aromatic molecules are bonded to a metal via a delocalized electron system.

Functional groups that interact with the chromophore are added to modify or enhance the color of the dye, make it more soluble in water, and attach it to the fibers of the cloth. By experimenting with different functional groups, it has been possible for chemists to generate dyes in a vast range of colors.

Synthetic dyes are also produced by adding other chemical groups – for example, nitro ($-NO_2$) groups, amino ($-NH_2$) groups, or halogens, such as fluorine, chlorine or bromine – to an

KEYWORDS

AROMATIC COMPOUND
AZO DYE
BOND
DYE
FASTNESS
FRIEDEL-CRAFTS REACTION
HYDROGEN BOND
MORDANT
PIGMENT

▷ At a leather tannery in Fez, Morocco, goatskins tanned with sumac (a local plant) are submerged in large tubs of red dye that give Moroccan leather its characteristic color. Dyeing is the final stage before the skins are made into fine leather goods.

NATURAL DYES

Extracts from plants and animals were used as natural coloring agents long before synthetic dyes became available. The the woad plant produces a blue dye **1**, and the purple dye Tyrian **2** is extracted from the shell of the marine mollusk *Murex*. The madder plant gave its name to a red-orange dye **3**; pollen of the purple autumn crocus is the basis for a bright yellow dye called saffron **4**.

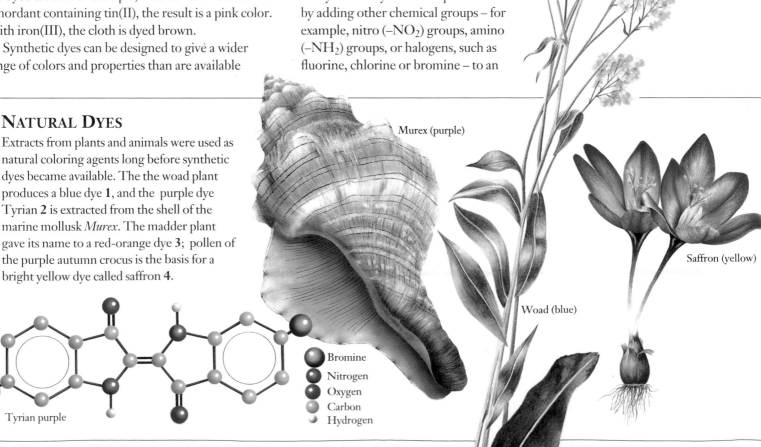

Murex (purple)

Saffron (yellow)

Woad (blue)

Tyrian purple

Bromine
Nitrogen
Oxygen
Carbon
Hydrogen

aromatic ring system, along with sulfonic acid. During this process, a series of chemical reactions known as the Friedel-Crafts reactions are often used to add carbon atoms to the aromatic rings and make it possible to build up side chains.

In the Friedel-Crafts reactions, aluminum chloride is used as a catalyst to help polarize halogen-containing organic molecules and cause them to substitute in a benzene ring. The reaction can be used to bring about both alkylation, in which an alkyl such as methyl ($-CH_3$) is substituted in, or in acylation, where an acyl group (RCO$-$, where R is a hydrocarbon group) is added. These reactions are also used widely in industry to synthesize hydrocarbons and other organic compounds, and in the manufacture of plastics such as polystyrene.

Besides being colorful, dyes must also be fast – that is, they must not fade in light, or after washing. Some dyes are made fast by the use of mordants; many azo dyes for cotton are fast because they are insoluble and become trapped in the fibers. Some dyes are held to the fibers by hydrogen bonding. But because hydrogen bonds are relatively weak, these dyes are only fast if their molecules are long and straight to allow them to line up with the fibers and form several hydrogen bonds.

The latest development is fiber-reactive dyes in which the dye is joined to the fibers by covalent bonds. These are available in a wide range of very bright colors and are extremely fast. The dyes can be modified easily to produce colors to order, and to attach to different types of fibers. As a result, color chemists can produce exactly the same color over a wide range of different types of fabrics.

Madder (red)

▷ Fast dyes (which do not fade when washed) are attached to cloth fibers by strong covalent bonds, instead of weak inter-molecular forces. To encourage the formation of covalent bonds it is necessary to modify dye molecules by adding groups which react chemically with the textile fibers. Here a colorfast dye for wool has been made by reacting an azo dye molecule contain-ing an amino (NH$_2$) group with a trichlorotriazine molecule 1. The reactive group that results 2 forms covalent bonds with amino groups in the wool fiber proteins to give fast dye 3.

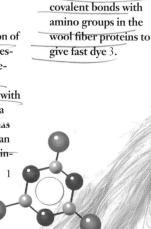

Dye

1

Trichlorotriazine

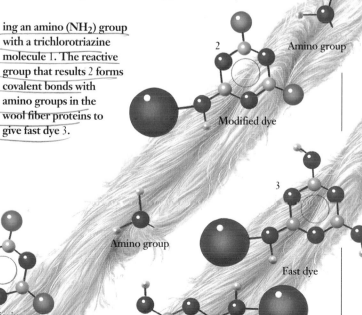

2

Amino group

Modified dye

Amino group

3

Fast dye

COSMETIC CHEMISTRY

THE pigments used as coloring agents by the fashion-conscious ancient Egyptians, Greeks and Romans would now be kept under lock and key as poisonous substances. For face makeup, white lead ($2PbCO_3Pb(OH)_2$) gave a pale color, red phosphorus was used as a rouge to add a touch of color to the cheeks and the mineral cinnabar (HgS) brightened the lips. Eyes were made more dramatic by the use of orpiment eye shadow (As_2S_3) and stibnite mascara

(Sb_2S_3). Because these cosmetics often contained metallic poisons such as lead, mercury, arsenic and antimony, cosmetics were often a health hazard as much as a beauty aid.

Today cosmetics are a good deal safer, and cosmetic companies run extensive tests to ensure that their products do not harm their customers. Modern cosmetics are made up of a relatively small number of substances, and the differences between brands are usually very slight in terms of their chemical composition.

Both face powders and eyeshadows are basically made up of pigments distributed in a base. Face powders generally contain an opaque substance such as zinc or titanium oxide (TiO) to cover the skin; mineral talc or zinc or magnesium stearate to provide adhesion and make the powder easy to apply; kaolin ($Al_4Si_4O_{10}(OH)_8$) or magnesium carbonate ($MgCO_3$) to absorb perspiration; and possibly guanine ($C_5H_5N_5O$) or mica ($KAl_2(AlSi_3O_{10})(OH)_2$) to give it sheen. To provide color, pigments are added, often as a coating on mica. To obtain a white color, titanium dioxide is used. Other colors can be obtained using pigments such as iron blue, carmine and iron oxide.

Lipsticks are made from mixtures of oily liquids such as castor oil; waxes, such as beeswax; and pigments.

▷ **Different colors in a cosmetic palette can be used for different effects. Pinks and reds add warmth and color; lavender can hide brown birthmarks, age spots and freckles; yellows enliven dark or sallow complexions.**

Good lipsticks provide a uniform intense color with good coverage, are shiny but not greasy, have a neutral taste, and are non-toxic and non-irritant. They are also formulated so that they do not melt in warm weather, or crumble in the cold. The lipstick itself is a mixture of castor oil and a wax such as beeswax or carnauba wax, which has a high melting point. It is designed to remain stiff in the tube but flow under pressure when being spread onto the lips. The color in lipsticks comes from dyes, often the same dyes as those used to color food. They include brilliant blue (a blue triphenylene dye), erythrosine (a red xanthene dye), amaranth (a red azo dye) and tartrazine (a yellow azo dye). For use in lipsticks, the water-soluble dyes are combined with aluminum oxide (AlO_3). This causes them to precipitate as an insoluble solid pigment, or lake. The lake is then suspended in castor oil, but does not actually dissolve in it.

In color-change lipsticks, which appear one color in the tube but change color when they reach the lips, a dye such as eosin (tetrabromofluorescein), which is lightly colored but turns red when it combines with the free amine ($-NH_2$) groups on the proteins in the skin, is added. The lipstick itself is usually colored by a lake dye. When it is spread on the lips, the lake dye is obscured by the eosin dye as it turns red.

▷ **Cosmetics are often used to highlight good features, to cover up scars or disfigurements, or to create a disguise. Some research has even shown that women wearing good makeup at a job interview are likely to be offered higher salaries.**

◁ Limited exposure to sunlight is good for the skin; overexposure can lead to premature aging and damage to the genetic material in skin cells. Excessive exposure can also cause skin cancer. Sunscreens, cosmetics used to protect the skin from sunlight, include compounds such as benzophenones and aminobenzoates. These absorb light in certain wavelength ranges and prevent it from reaching the skin. To be effective, sunscreens must remain chemically stable in light. They must also be soluble in the cosmetic base to make them easy to apply, but insoluble in water or perspiration so that they do not easily wash off. Sunscreens have become available in a wide range of sun protection factor (SPF) ratings.

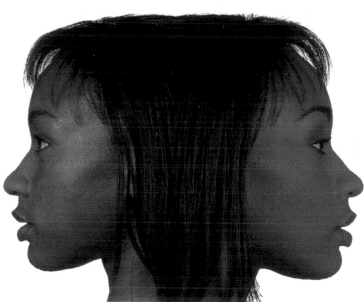

☐ Two different women as they appear with and without makeup. Foundations are used to cover wrinkles or even out skin tone, make the skin appear younger, and to provide a good base for other types of makeup.

Well chosen eye makeup (shadow, eyeliner and mascara) enhances attractive features and makes the face look more lively. Blushers can be used to add color and a healthy glow to the skin, as well as for contouring the shape of

the face to bring out high cheekbones, or to make low cheekbones look higher. Lip colors not only protect and moisturize the lips; they also brighten the face by adding color. Skin-colored powder "finishes" the surface and the look.

PHOTOGRAPHY

THERE are few more graphic ways to demonstrate the interaction between light and chemicals than to take a picture of it. The energy associated with photons of light causes chemical reactions in some substances. Photography is based on the effect of light on silver halides, the salts in the emulsion coating the photographic film. A photographic emulsion is made up of grains of a silver halide, which are suspended in gelatin and used as a coating for a plastic strip, which

acts as a support. The most commonly used silver halide is silver bromide (AgBr), but silver chloride (AgCl) and silver iodide (AgI) are also used. Silver chloride reacts more slowly to light than silver bromide, while silver iodide reacts more quickly.

When a photograph is taken, the film is exposed to light for a fraction of a second. This causes some of the millions of molecules in each grain to break down into silver atoms and bromide ions. The greater the exposure to light, the more molecules are broken down. Thus, the area of the film exposed to white or light-colored objects contains a number of silver atoms resulting from the breakdown of silver bromide, but the area exposed to darker objects does not contain any.

This latent image is transformed into a picture by using a strong reducing agent, such as hydroquinone ($C_6H_4(OH)_2$), as a developer. The few atoms of silver in the areas of the film exposed to light-colored objects act as catalysts of a reaction in which the developer reacts with the silver halide, causing it to break down into silver and halogen ions. The areas of the film exposed to dark-colored objects have no silver atoms to act as a catalyst, so no silver halide is broken down. In gray areas, only a few silver halide molecules are broken down.

The result of the developing process is a negative image. In this image, black and white are reversed, because the silver halide is white, whereas the silver atoms appear black. The negative is fixed by using a soluble salt such as sodium thiosulfate ($Na_2S_2O_3$). This reacts with the silver halide grains to prevent them from decomposing in the light and forming silver deposits which would turn the photograph black.

A positive print is made by placing a piece of paper coated with silver halides behind the negative and exposing the negative to light. The darker areas of the

▷ Creative photography does not end with releasing the camera shutter. Many important changes to the picture can be made while making a print in the darkroom. The photographer can use techniques such as "dodging" and "burning in". A mask may be held above the surface of the printing paper under the enlarger to hold back light reaching one part of the picture (such as the silhouette of the flying bird) or to allow extra light to reach a bright area (such as the sky on the right of the picture). The print may also be toned with selenium to alter the contrast between dark and light areas.

▣ An early kind of photography was invented by the Frenchman Louis Daguerre in the 1830s. In making a daguerrotype, a silver-coated copper plate was sensitized in iodine vapor and then exposed to light inside a camera. The plate was then developed and fixed to produce a reversed black image RIGHT. At the turn of the century, many prints were made in sepia tone ABOVE.

negative prevent light from getting through to the paper and causing the silver halide to decompose, whereas the lighter areas allow the light through. When the paper is developed in a similar way to the film, a positive image results.

Color photography is based on similar principles. The film consists of silver halide in gelatin, but it is overlaid with three layers which contain precursors of azo dyes, separated by filters which only allow light of a certain wavelength through. Only three different dyes are needed in color film because our eyes contain three receptors which respond to blue, green and red. The vast range of colors we see results from a mixture of these three primary colors. The dyes used in color film include a cyan dye, which absorbs red light and reflects blue and green; a magenta dye, which absorbs green light; and a yellow dye which absorbs blue light.

When the film is exposed to light, each dye reacts to specific wavelengths, and allows those wavelengths to pass on to the silver halide grains. This causes some of the grains to decompose. As the light penetrates through the successive layers, a negative image in each color range is formed. When the film is printed, the colors are reversed to give a positive picture in color.

▽ An exposed film is developed by placing it, in the dark, into a developing tank 1, then agitating it in a developer 2 for a specific time, during which a chemical reaction produces tiny grains of metallic silver. After development, the film is fixed 3 to remove any unexposed silver salts; excess chemicals are washed away with water. The chemical changes are shown in the circles, those in exposed parts of the film on the left, and in unexposed parts on the right.

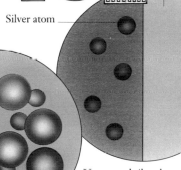

Silver bromide removed

Silver atom

Silver atom

Silver bromide

Unexposed silver bromide

Silver ion

Silver bromide

Bromide ion

△ Similar photochemical reactions are used to make an enlarged positive print on silver halide-coated paper from the negative film. Light converts a small amount of the silver ions in the emulsion to metallic silver – only a few atoms among many millions. The chemical process of developing completes the conversion. Any grains not exposed to light remain as silver halides and are removed by fixing and washing. The areas exposed to most light have the most silver grains and appear darkest on the print.

PHOTOSYNTHESIS

Green plants use photosynthesis to capture energy radiated from the Sun. This sustains all life on Earth. In photosynthesis, water molecules are split and combined with carbon (derived from carbon dioxide in the atmosphere) to make the sugar glucose. The glucose is stored in the form of its polymer starch. It may be used to make the straight-chain polymer cellulose (the major supporting material in plant cell walls) or broken down by the plant during respiration to release energy. Most of the oxygen in the atmosphere that animals breathe is a byproduct of this reaction.

The key molecules in all light-driven biochemical reactions are biological pigments, which capture the energy of light when incoming photons boost the electrons in some of the pigments molecule's atoms to a higher energy level. The key pigment is chlorophyll, a porphyrin that has a magnesium (Mg^{2+}) ion at its center. Several small side-chains, attached outside the porphyrin ring, alter the absorption properties in different types of chlorophyll.

Porphyrins are derivatives of porphin, a simpler purple compound made up of pyrroles (containing carbon, nitrogen and hydrogen atoms), joined into a ring by methylene (–CH=) groups. Porphyrins readily lose their central hydrogen atoms to take on a negative charge. The charge is neutralized by positively-charged metal ions such as iron (Fe^{2+}), magnesium (Mg^{2+}) and cobalt (Co^{2+}), which fit into the center of the porphyrin molecule. Other well-known porphyrins include hemoglobin, the oxygen-carrying protein in blood (which has an Fe^{2+} ion at its center), and vitamin B12, which helps to synthesize amino acids (it has Co^{2+} at its center).

Chlorophyll absorbs light energy in the red and blue region of the visible spectrum, and thus appears green. During photosynthesis, it transfers this light energy into chemical energy. This happens when the photons of light absorbed by the chlorophyll excite the electrons of the magnesium ions. The electrons are then channeled away through the carbon bond system of the porphyrin ring to fuel photosynthesis.

Photosynthesis involves three series of chemical events: the light reactions and the dark reactions, during which energy is captured and stored; and a series of reactions to replenish the pigment.

The light reactions can take place only in the presence of light and occur on photosynthetic membranes in the chloroplasts of plants. During the reactions, a photon of light is captured by the chlorophyll molecule and excites an electron within the pigment. The excited electron travels along a series of electron-carrier molecules in the photosynthetic membrane to a transmembrane proton-pumping channel, where it induces a proton to cross the membrane. The proton later crosses back across the membrane, which drives the synthesis of the energy-carrying molecule adenosine triphosphate (ATP). In addition, a second type of energy-carrying molecule, nicotine adenine dinucleotide phosphate (NADP), is reduced to form the electron carrier NADPH.

During the dark reactions, the energy from ATP and NADPH is used to make organic molecules from atmospheric carbon dioxide (CO_2) in a cycle of enzyme-catalyzed reactions known as carbon fixation.

During a third series of reactions, the electron that was stripped from the chlorophyll at the beginning of the light reactions is replaced. Without this, the continuous removal of electrons from chlorophyll in photosynthesis would cause it to become deficient in electrons and it would no longer be able to trap photon energy by electron excitation.

▷ **Photosynthesis involves three series of biochemical reactions reactions. These include light and dark reactions, during which the energy of light is captured and – through the mediation of the pigment chlorophyll – used to combine carbon dioxide and water to form sugars. (This process is called "fixing" carbon.) This conversion takes place in special tiny structures in the plant cells, called chloroplasts. Light reactions require light and take place on the thylakoid membrane – formed by flat, round sacs in the chloroplast. Dark reactions occur in the stroma, a thick fluid within the chloroplast, enclosed by a separate membrane. In the third series of reactions in photosynthesis, chlorophyll is replenished with electrons, ready to begin the cycle again.**

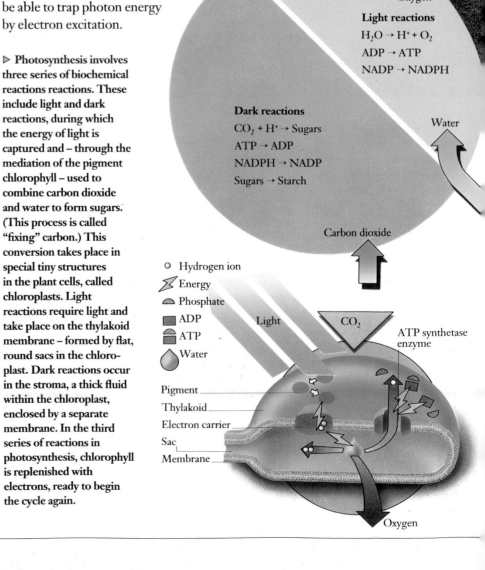

Oxygen

Light reactions
$H_2O \rightarrow H^+ + O_2$
$ADP \rightarrow ATP$
$NADP \rightarrow NADPH$

Dark reactions
$CO_2 + H^+ \rightarrow$ Sugars
$ATP \rightarrow ADP$
$NADPH \rightarrow NADP$
Sugars $\rightarrow$ Starch

Water

Carbon dioxide

○ Hydrogen ion
⚡ Energy
◠ Phosphate
▭ ADP
▱ ATP
⬠ Water

Light

CO_2

ATP synthetase enzyme

Pigment
Thylakoid
Electron carrier
Sac
Membrane

Oxygen

■ Green plants capture
around 1 percent of the
energy radiated from the
Sun to the Earth. This is
the source of all energy for
life on Earth. More than
150 billion tonnes of sugars
are produced yearly by
photosynthesis.

■ Chlorophyll,
the magic ingredient that
makes photosynthesis
possible, is contained in the
chloroplasts below the
outer skin of the leaves.
Tiny holes, called stomata,
on the underside allow gas
exchange between the leaf
and the air.

Cuticle
Epidermal cell

Phloem sieve tubes
Stoma

Sugar
Chloroplast
Mesophyl
Water

Glucose passes
down to roots,
for growth

Water passes
up trunk and
branches

■ Disk-shaped thylakoids
are the chemical factories of
a cell, and they are the site
at which photosynthesis
takes place. Chemical
energy from sunlight is
held in molecules of ATP
and NADP, which power
the dark reactions. In the
thylakoid, water is split to
produce oxygen and
hydrogen ions, which
combine with NADP to

form NADPH. The oxygen
and hydrogen ions also pass
through the membrane and
convert ADP into ATP.
The energy of NADPH
and ATP powers the
synthesis of sugars.

ATMOSPHERIC CHEMISTRY

THE Earth is unique among the planets of the Solar System in having a chemically reactive atmosphere rich in oxygen. The chemistry of the atmosphere is a series of delicately balanced cycles involving many chemicals that interact closely. Photochemical reactions driven by the energy in sunlight play a key role.

Without the atmosphere, the surface of the Earth would have an average temperature of around -18°C.

The greenhouse effect traps some of the energy of the Sun, raising the surface temperature to a more comfortable 15°C. This effect occurs because the "greenhouse gases", water vapor and carbon dioxide (CO_2), trap radiated heat, which originates as high-energy short-wavelength radiation from the Sun. This energy is absorbed by the surface of the Earth which, in turn, emits radiation of its own, but at much longer and lower-energy infrared wavelengths. Some of this radiated heat is trapped in the lower part of the atmosphere (the troposphere) by water vapor and carbon dioxide. As in a greenhouse, the air near the surface of the Earth is warmed because the greenhouse gases do not allow all of the radiation to escape.

The burning of fossil fuels raises the levels of CO_2, upsets the natural balance of greenhouse gases in the atmosphere and strengthens the greenhouse effect. Other gases released by human activities, including ozone, methane, nitrogen oxides and chlorofluoro-carbons (CFCs), also contribute to the greenhouse effect because they absorb radiation in the 7–13 micrometer wavelength range, a window through which 70 percent of the radiation from the Earth's surface normally escapes into space. The radiation window occurs because the two main greenhouse gases, water vapor and carbon dioxide, absorb radiation in the 4–7 and 13–19 micrometer wavelength bands respectively.

CFCs also cause problems in the upper part of the atmosphere (the stratosphere). There, ozone (O_3) plays an important role in filtering out dangerous ultraviolet radiation from the Sun and preventing it from reaching the surface of the Earth. It does this by absorbing ultraviolet radiation in the 4–400 nanometer wavelength band as it splits to form oxygen molecules and free oxygen atoms ($O_2 + O$). The protective ozone layer forms when oxygen (O_2) in the

KEYWORDS

CHLOROFLUOROCARBON (CFC)

FREE RADICAL

GREENHOUSE EFFECT

NITROGEN OXIDE

OZONE HOLE

PHOTOCHEMISTRY

PHOTODISSOCIATION

SULFUR DIOXIDE

▨ **Delicately balanced photochemical reactions** RIGHT **play a key role in the Earth's atmospheric chemistry. In previous centuries, there was very little atmospheric pollution. Since the mid-20th century, increasing amounts of carbon dioxide (CO_2) in the atmosphere are trapping more heat around the Earth. This "greenhouse effect" prevents the re-emission of the Sun's radiation back into space and may be the cause of global warming.** BELOW **Sulfur dioxide, nitrogen oxide and other industrial emissions oxidize in clouds to form acid rain, which damages forests, lakes and soil. Near the ground, nitrogen oxides and carbon monoxide from vehicle exhausts combine with low-level ozone to form photochemical smog, visibly polluting the air.**

stratosphere absorbs particular frequencies of ultra-violet radiation from the Sun. This causes some of the O_2 molecules to photodissociate – split up into single oxygen atoms. The oxygen atoms are free radicals, bits of broken-down molecules that are unstable and highly reactive. Some of these oxygen radicals react with the remaining O_2 molecules to form ozone (O_3).

The ozone itself absorbs ultraviolet radiation and photodissociates naturally. Left to themselves, the reactions that cause the production and destruction of stratospheric ozone would reach a steady state in which as much ozone was produced as was destroyed. However, various chemicals known as free radicals, including chlorine atoms, hydroxyl radicals (HO) and nitrogen oxides (NO_x), upset the balance because they speed up the removal of ozone from the stratosphere by reacting with it.

CFCs act as a vehicle to bring chlorine free radicals into the upper atmosphere. CFCs are extremely stable and remain intact until they reach the stratosphere. There they are broken down by the fierce ultraviolet radiation to release free chlorine atoms. Because each chlorine atom can destroy roughly 100,000 molecules of ozone, the protective stratospheric ozone layer is under considerable threat. In 1985, scientists at the British Antarctic Survey recognized a seasonal

Al^{3+}	Aluminium ion
CH_4	Methane
C_2H_4	Ethene
CO_2	Carbon dioxide
H^+	Hydrogen ion
HNO_3	Nitric acid
H_2SO_4	Sulfuric acid
NH_3	Ammonia
NH_4	Ammonium ion
NO	Nitric oxide
NO_2	Nitrogen dioxide
O_3	Ozone
SO_2	Sulfur dioxide
SO_4^{2-}	Sulfate ion

Reflected

Incoming solar radiation

Absorbed

Reflected

Reemitted

Level of stratosphere

Absorbed

Reflected

Absorbed

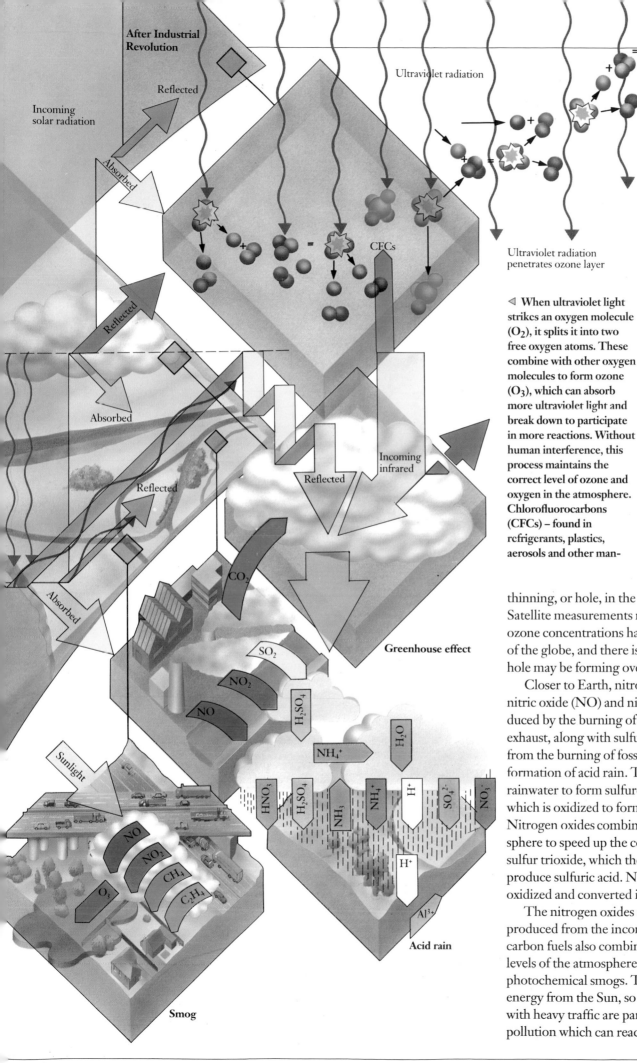

After Industrial Revolution

Reflected

Incoming solar radiation

Absorbed

Reflected

Absorbed

Reflected

Absorbed

CO₂

Incoming infrared

Reflected

CFCs

Ultraviolet radiation

Ultraviolet radiation penetrates ozone layer

Oxygen atom
Oxygen molecule
Ozone molecule
CFC molecule
Chlorine monoxide

Greenhouse effect

SO₂

NO₂

NO

H₂SO₄

NH₄⁺

H₂O

HNO₃

H₂SO₄

NH₃

NH₄⁺

H⁺

SO₄²⁻

NO₃⁻

H⁺

Al³⁺

Acid rain

Sunlight

NO

NO₂

CH₄

O₃

C₂H₄

Smog

◁ **When ultraviolet light strikes an oxygen molecule (O_2), it splits it into two free oxygen atoms. These combine with other oxygen molecules to form ozone (O_3), which can absorb more ultraviolet light and break down to participate in more reactions. Without human interference, this process maintains the correct level of ozone and oxygen in the atmosphere. Chlorofluorocarbons (CFCs) – found in refrigerants, plastics, aerosols and other man-made products – rise into the atmosphere and lose chlorine atoms when struck by sunlight. These "free radical" chlorine atoms destroy ozone molecules by taking away oxygen atoms from them. Chlorine free radicals are not only hugely destructive, they also remain in the stratosphere for up to 130 years. Massive loss of ozone results in a "hole" in the layer over the Antarctic which has been increasing in size by up to 50 percent annually since 1979.**

thinning, or hole, in the ozone layer over Antarctica. Satellite measurements now show that stratospheric ozone concentrations have decreased over other parts of the globe, and there is evidence that a smaller ozone hole may be forming over the Arctic.

Closer to Earth, nitrogen oxides, which include nitric oxide (NO) and nitrogen dioxide (NO_2) produced by the burning of fossil fuels and emitted in car exhaust, along with sulfur dioxide (SO_2) produced from the burning of fossil fuels, contribute to the formation of acid rain. The sulfur dioxide reacts with rainwater to form sulfurous acid (H_2SO_3), some of which is oxidized to form sulfuric acid (H_2SO_4). Nitrogen oxides combine with ozone in the atmosphere to speed up the conversion of sulfur dioxide to sulfur trioxide, which then dissolves in water to produce sulfuric acid. Nitrogen dioxide (NO_2) is itself oxidized and converted into nitric acid (HNO_3).

The nitrogen oxides and carbon monoxide produced from the incomplete burning of hydrocarbon fuels also combine with ozone in the lower levels of the atmosphere to pollute the air and form photochemical smogs. The reaction is speeded up by energy from the Sun, so that in sunny climates cities with heavy traffic are particularly prone to air pollution which can reach dangerous levels.

CHEMICAL
Analysis

NEARLY ALL OTHER BRANCHES of chemistry rely on analytical techniques – one of the keystones of chemical science. Chemical analysis has many applications. It ensures that food and drink contain what we think they do. Analytical chemistry is also an important tool for diagnosing disease, because many medical disorders lead to changes in the concentrations and production of body chemicals. Analytical techniques help unravel crimes by providing vital information about the fuel used to start a fire, or the presence or absence of drugs, blood or other body fluids. Within the chemical industry itself, analytical chemistry provides the means to monitor the purity of the products made.

Analytical chemists are concerned with two basic questions: What? and How much? By qualitative analysis, chemists can identify the elements present in a compound or mixture of substances, even if they are present only in tiny amounts. Quantitative analysis is used to determine how much of a component is present. Techniques of analytical chemistry range from simple qualitative analysis using reagents (chemicals that help to identify other substances by the reactions they cause) or determining melting or boiling point, to complex techniques such as mass spectrometry, chromatography and nuclear magnetic resonance spectroscopy.

Analysis is an important part of any chemist's job, and chemists rely on a wide range of techniques. Some involve mainly weighing, measuring and identifying substances based on their reactions with other chemical substances (reagents). Others, such as spectroscopy, chromatography and nuclear magnetic spectroscopy, are highly sophisticated and require expensive equipment. These methods make it possible to analyze very small amounts of compounds with a high degree of accuracy. But all techniques require the basic skills of a chemist to prepare samples and interpret the results.

FORENSIC CHEMISTRY

AT THE SCENE of a crime, every contact leaves a trace. Working according to this motto, forensic chemists use such traces to provide the police with information to assist in solving the crime. This information is also used to help the courts decide if a crime has been committed, and if so, by whom.

Forensic chemistry requires excellent analytical and interpretive skills combined with a creative mind to make it possible to come up with the best methods of analysis to solve the problem at hand. The conditions under which the scientists work can be difficult, and the nature of the samples they analyze is often unusual. For example, forensic chemists may need to analyze very small amounts of unknown substances which may be preserved in other substances such as blood, urine, human tissue or stomach contents. Often the material being analyzed is in very poor condition because of damage due to fire, putrefaction or deliberate attempts at concealment.

KEYWORDS

BLOOD TEST

BREATH TEST

CHROMATOGRAPHY

DRUG

GAS-LIQUID
  CHROMATOGRAPHY

QUALITATIVE ANALYSIS

QUANTITATIVE ANALYSIS

SPECTROSCOPE

URINE TEST

The range of materials forensic chemists routinely detect is vast. Aside from blood, body fluids and poisons – the classic clues familiar to all readers of detective novels – modern forensic chemists also test for flakes of paint, splinters of glass, drugs, traces of gunpowder left by firearms, and anything else that has a distinctive chemical signature.

Their powers of discrimination are great. Knowing that every brand of gasoline is made up of varying proportions of different hydrocarbon fractions, forensic chemists are even able to determine whether a fire in a gas station occurred accidentally or due to arson. If the chemical signature of all the gasoline traces matches the brand of gasoline sold at that location, the fire may have started accidentally. But if an indication of another brand can be found, arson is the more likely cause.

In their work, forensic chemists draw on techniques familiar to all analytical chemists. The various forms of chromatography and spectography are particularly useful for the identification of traces of substances ranging from dyes and cigarette ash to drugs and poisons. Because the substance to be examined may consist of thousands of samples of tiny quantities, many of the analytical techniques employed are highly automated to allow samples to be processed in large batches rather than one at a time.

Forensic chemists also make use of powerful molecular biological and biochemical techniques, such as the polymerase chain reaction (PCR) and DNA fingerprinting, to identify suspects and victims by the traces of body fluids, skin or hair they leave behind.

Immunoassay techniques can be used to detect traces of poison long after a murder has been committed – months or even years later. Immunoassays rely on biological molecules (known as antibodies) to recognize and bind to specific proteins (known as antigens). An antibody fits with a particular antigen like a key fitting into a lock. With immunoassays it is possible to detect, identify, measure or purify almost any substance against which specific antibodies can be raised. The assays can detect tiny amounts of the target substance, but it is necessary to test for each type of substance. In radioimmunoassays, radioactively labeled antigens representing the substance in question are added to a sample. If antibodies to the substance are present, they attach to the labeled antigens. After the antibody-antigen complexes are separated out, the radioactivity of the solution containing the complexes is measured. This gives an indication of whether the antibodies present in the sample have attached to the labeled antigens – and thus whether the substance is present.

■ In a true case, forensic chemists from the London Metropolitan Police Forensic Laboratory solved a murder. Routine samples taken from the body of a landscape gardener were studied by medical students. One student noticed unusual traces. Forensic chemists conducted radioimmunoassays and high-performance liquid chromatography on the year-old samples, which had been preserved in formalin. The forensic chemists were able to detect the presence of paraquat, a poisonous water-soluble ammonium compound used in some weedkillers. The gardener's widow admitted adding weedkiller to her husband's drink. She was convicted of murder.

ONE YEAR LATER...

VOLUMETRIC ANALYSIS

THERE are two types of chemical analysis. A chemist may wish to determine the composition of a substance – to find out what chemicals it contains. This is called qualitative analysis. Alternatively, the chemist may know the identity of a substance but need to determine what its concentration is. This is the purpose of quantitative analysis.

One of the simplest ways of determining how much of a compound is present is by weighing it. Weighing,

or gravimetric analysis, is one of the most common and fundamental procedures in quantitative chemical analysis. It works by taking a known weight of the substance to be analyzed, changing it to another known substance by means of chemical reactions, and then reweighing the products. From a knowledge of the weights and the reactions, a chemist can work out the purity of the original substance.

A typical gravimetric analysis involves three main steps. Firstly, the weighed sample is dissolved to give a solution. Next the element or ion being analyzed is reacted with a selective reagent to precipitate it out of the solution as one of its compounds. Finally the precipitate is filtered, dried, and sometimes reacted further or strongly heated (to form a more stable compound) before being weighed.

Gravimetric analysis is normally carried out on small amounts of substances and can give an accuracy of better than 1 percent. The method depends on the use of very accurate chemical balances, the laboratory instruments that are employed for the precise measurement of small masses. There are many types, including top loading, triple beam, equal arm, single pan substitution, hybrid balances, electromagnetic force balances and electronic balances – which may measure in units as small as 0.1 microgram (10^{-7} grams). Most electronic balances can be connected with printers, computers and other specialized application devices.

Volumetric analysis involves reacting chemicals in solution, usually to determine the strength of one of them. For example, if a chemist wants to find out the strength of a solution of an acid, he or she reacts it with an alkali of known strength. The alkali is made up by accurately weighing an amount of the solid substance and dissolving it to make a precise volume of solution – known as a standard solution. Some of the standard

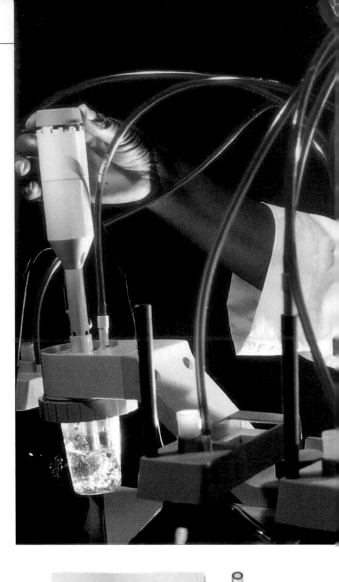

▷ A scientist adjusts an automatic titration device used for quality control in drug manufacture. In the first stage of titration, a standard solution of known concentration is prepared by weighing out a compound, dissolving it in water, and then adding water to make up the solution to a known volume. In stage two, the standard solution is placed in a burette, a precisely calibrated volume measuring vessel. A measured amount of a solution of unknown concentration is placed in a flask. In the third stage, the standard solution is carefully added, drop by drop, until the end point (when the solutions have reacted completely) is reached. The volume of standard solution is used to calculate the concentration of the solution being tested.

1 Weighed sample dissolved in known volume of solution (standard solution).

2 Sample (unweighed) dissolved in water and known volume removed with pipette.

3 Standard solution run into known volume from burette until reaction is complete.

solution is used to fill a burette. Using a pipette, the chemist takes a precise volume of the acid solution and adds it to a flask. Then the standard solution is run out of the burette into the flask until the alkali exactly neutralizes the acid. From a knowledge of the reacting volumes and the strength of the standard solution, the chemist can work out the strength of the acid. The whole technique is known as a titration.

The method depends on the chemist's knowing exactly when the neutralization reaction is complete. To do this the chemist adds an indicator to the liquids in the flask. In this example, an indicator such as litmus, which changes color from red to blue when the solution just ceases to be acid and is becoming alkaline, could be used.

Another acid–alkali indicator is phenolphthalein, which is colorless in acid solutions but turns bright red in alkali. Acid–alkali titrations are widely used in the soft drinks industry to check the acidity of the product, and for such large-scale analyses the procedure is usually automated.

Other types of titrations require different indicators. For example, the presence of iodine in solution is revealed by using starch as an indicator; it forms a deep blue color in the presence of iodine.

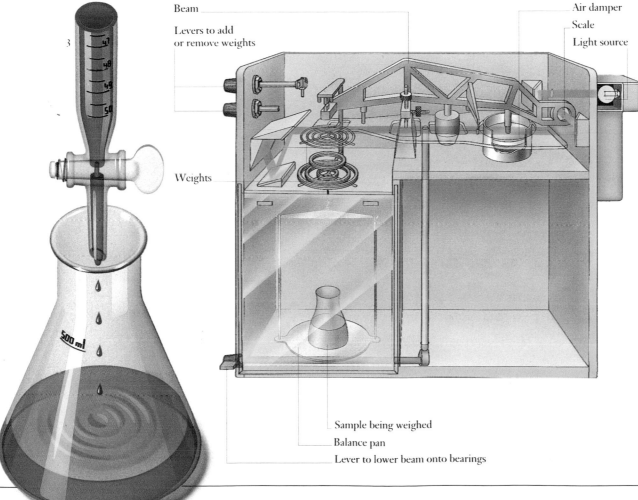

Beam

Levers to add or remove weights

Weights

Air damper

Scale

Light source

Sample being weighed

Balance pan

Lever to lower beam onto bearings

◁ **Accurate weighing is the basis of gravimetric analysis. Precision chemical balances are an essential tool. There are many different types in use, each designed to work at different levels of precision. In general, the more precise the balance, the smaller the amount of substance that can be weighed on it at one time. Air damp balances are popular because, while reasonably accurate, they make use of air resistance to slow down the oscillation of the beam, and thus speed up the weighing process. In the type illustrated, the object being weighed is counterbalanced by circular weights added by means of levers and controlled by the knobs on the left. The final decimal place of the weight is given on a graduated scale, projected from the balance beam.**

CHROMATOGRAPHIC ANALYSIS

SOME methods of chemical analysis rely on separating a mixture of substances and then identifying the components. Chromatography, by contrast, is a method of separating several substances on the basis of their differences in solubility. The separation occurs because compounds pass through a so-called stationary phase at different rates depending on their molecular characteristics. The compounds being tested are in a mobile phase, which may be either gas or liquid. The stationary phase is generally a solid or a liquid on a solid support.

In some forms of the technique, the separated components appear as colored bands or spots on paper, which gives chromatography its name, though chromatography can also be used to separate colorless substances. After the chromatogram is run, it can be developed by spraying it with a chemical known as a locating agent, which attaches to the spots of separated components and makes them appear colored.

There are many different forms of chromatography, including column

chromatography (the earliest form); thin-layer chromatography; paper-liquid chromatography; gas-liquid, or gas, chromatography; and high-performance liquid chromatography. All of these forms of chromatography rely on the same basic principle: differences in solubility between different molecules. These differences depend on their chemical nature and the size of the molecules.

In all forms of chromatography the substance to be analyzed is dissolved in a liquid or a gas. This is known as the mobile phase. To carry out the analysis, this solution is passed through a stationary phase, which may be a packed column, a solid surface or a liquid on a solid support. The stationary phase acts as a molecular obstacle course for the mobile phase. As the different substances dissolved in the mobile phase pass through the stationary phase, some are more attracted to the stationary phase, and thus move more slowly. Others remain more attracted to the mobile phase, and these move more quickly. The substances can be separated on the basis of their different rates of movement, and they can be identified by measuring the rates at which they move under specific conditions.

▼ Each component of the sample travels through the column at a different rate, depending on whether it is more attracted to the mobile or to the stationary phase. Those favoring the mobile phase are carried more quickly. A detector at the end of the column monitors the components as they emerge from it. Signals from the detector are plotted out by a recorder to produce a chromatograph. This takes the form of a series of peaks which represent the concentrations of the various components in the sample.

▷ In gas-liquid chromatography (GLC), the mobile phase is an unreactive gas such as nitrogen. The stationary phase is made up of a small amount of liquid adsorbed on the surface of the particles of a fine powder packed into a long thin column coiled in an oven. The sample is injected into the carrier gas stream just before it enters the chromatograph column.

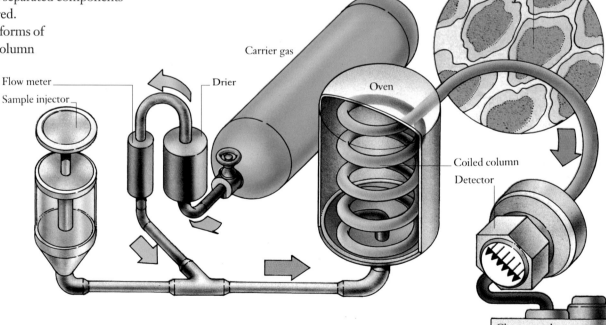

Packing in column

Flow meter

Sample injector

Drier

Carrier gas

Oven

Coiled column

Detector

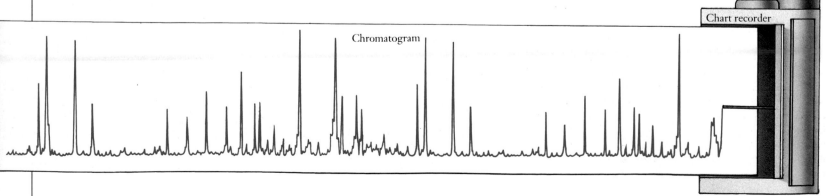

Chart recorder

Chromatogram

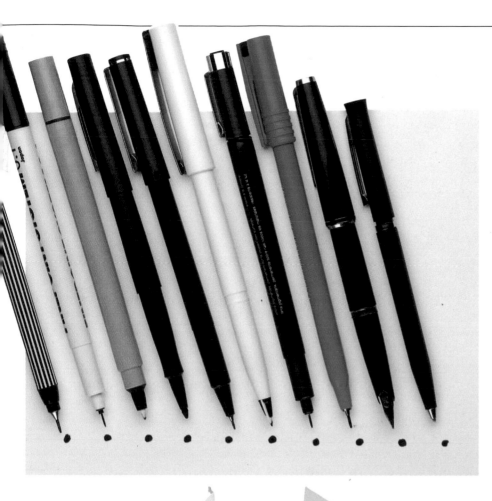

Chromatography is a useful laboratory technique for separating, identifying and quantifying the number of molecules or elements in a solution, all at the same time. In industry it is used for testing for purity, or identifying unknown substances by comparing them with known components. Gas-liquid chromatography and, increasingly, high-performance liquid chromatography provide a quick method for forensic scientists to test for the traces of explosives, or for sports officials to test for the presence of illegal drugs such as steroids in the urine of athletes.

Most forms of chromatography can be used only to separate very small amounts of components, but column chromatography provides a useful method for separating out large quantities of materials, and for this reason it is often used in larger-scale industrial separation operations. In this technique, an absorbing substance such as alumina (aluminum oxide) is packed into a vertical glass tube. A solution of the mixture is poured into the column and then washed through continuously with a solvent – a process called elution. Different substances travel down the column at different speeds because they are absorbed to different extents. The liquid that emerges, known as an eluate, is collected in "fractions", each of which contains one of the substances in the mixture. A similar method is ion-exchange chromatography, in which ions from an ion-exchange resin are displaced by the eluant.

■ All black inks look similar, but they are actually made up of different mixtures of pigments. Paper chromatography is a simple technique that can be used to tell them apart.

To carry out the analysis, samples of the black ink to be tested are spotted near the edge of a piece of blotting paper. The paper is rolled into a cylinder and placed, dots down, into a beaker containing a solvent. The solvent climbs up the paper by capillary action.

As the solvent rises up the paper, the various components are carried to different distances, depending on their relative solubilities in the mobile and stationary phase. The differences between two inks may be quite dramatic.

The distance that is traveled by each component is described by its value relative to the front, or Rf value, which is defined as the distance moved by the colored spot divided by the distance moved by the solvent front. The exact composition of the different inks can then be determined by comparison with various substances whose Rf values are known.

SPECTROSCOPIC ANALYSIS

ANALYTICAL chemists in laboratories have something in common with astronomers who observe the sky through a giant telescope: both use spectroscopy to discover the chemical composition of the objects they study. The techniques ultimately rely on the absorption or emission of energy by atoms or molecules, whether in a test tube or in a remote star.

When an element, such as a metal, is heated to a very high temperature it becomes incandescent – it gives off light. If the light is passed through a prism, it is split into its component wavelengths (colors) which form a spectrum consisting of a series of lines. When a high-voltage electric current is passed through a sealed tube containing gas at low pressure, the gas glows with light that can also form a line spectrum.

The line spectra from incandescent solids and gases are examples of emission spectra. They occur because outer electrons in the atoms concerned absorb energy (thermal or electrical) and become "excited", so that they temporarily occupy high energy levels. Then, when the excited electrons "jump" back to their original levels, they emit their excess energy in the form of light. The wavelengths of the emitted light are characteristic of the electron jumps and may be used to identify the element.

Other major forms of spectroscopy involve absorption spectra, which rely on the measurement of wavelengths absorbed by molecules and atoms. In infrared spectroscopy, a sample of a substance (often dissolved in a suitable solvent) is irradiated with infrared radiation. In an organic compound, radiation is absorbed by the chemical bonds between various atoms within its molecules. (The extra energy makes the bonds vibrate more vigorously.) Detectors measure the wavelengths as they are absorbed to produce an infrared absorption spectrum of the substance, from which it can be identified.

In ultraviolet absorption spectroscopy, using ultraviolet radiation, the radiation is absorbed by outer electrons of atoms in the substance being analyzed which become excited to higher energy levels. The absorbed wavelengths are measured, to identify the atoms concerned. The technique is suitable for both inorganic and organic substances.

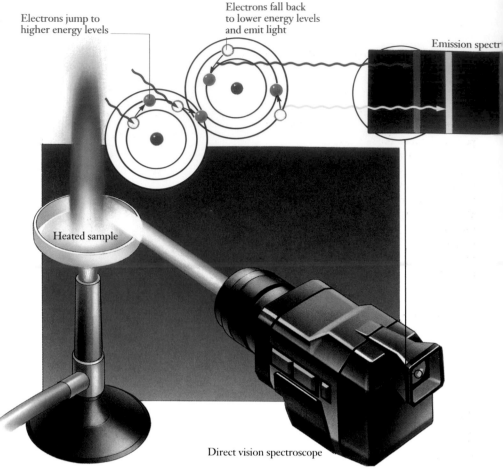

Electrons jump to higher energy levels

Electrons fall back to lower energy levels and emit light

Emission spectr

Heated sample

Direct vision spectroscope

Spectra can even be produced by studying the absorption of energy by the nuclei of atoms. The stimulating radiation consists of radio waves, and the technique is known as nuclear magnetic resonance (NMR) spectroscopy. The protons in an atomic nucleus spin, acting as miniature magnets. In the presence of a powerful external magnetic field, these magnets line up with or against the field. Their energy levels can be changed by irradiating a sample with microwaves, which are momentarily absorbed. The frequency of the absorbed energy can be used to identify particular atoms. Hydrogen, fluorine and phosphorus (with odd numbers of protons) exhibit this behavior. But the presence of hydrogen in virtually all organic compounds makes NMR a powerful tool in organic analysis. It has been developed as a non-surgical method of studying living tissues. Electron spin resonance (ESR) spectroscopy is a similar technique based on electron spin transitions in atoms with an unpaired electron.

A mass spectrometer separates a mixture of atoms according to their different masses. The sample is converted into positive ions and passed (in a vacuum) between the poles of a magnet. The ions are deflected by the magnetic field; the heavier ions are deflected less less than the lighter ones. The separated ions are then recorded on a photographic plate or by a series of detectors, producing a mass spectrum.

△ A simple test for identifying certain metallic elements involves heating them to incandescence in a Bunsen burner flame and noting any characteristic color produced. A more sophisticated version of this test employs a direct-vision spectrometer to reveal colored lines in an element's emission spectrum. The light is produced when electrons, initially excited to a higher energy level by the applied heat, jump back to their normal, lower energy levels. The technique was introduced by the German chemist Robert Bunsen who in 1860 and 1861, with his assistant Gustav Kirchhoff, discovered the new elements cesium and rubidium, naming them for the colors of their spectrum. *Cesium* is Latin for blue-gray and *rubidius* means red.

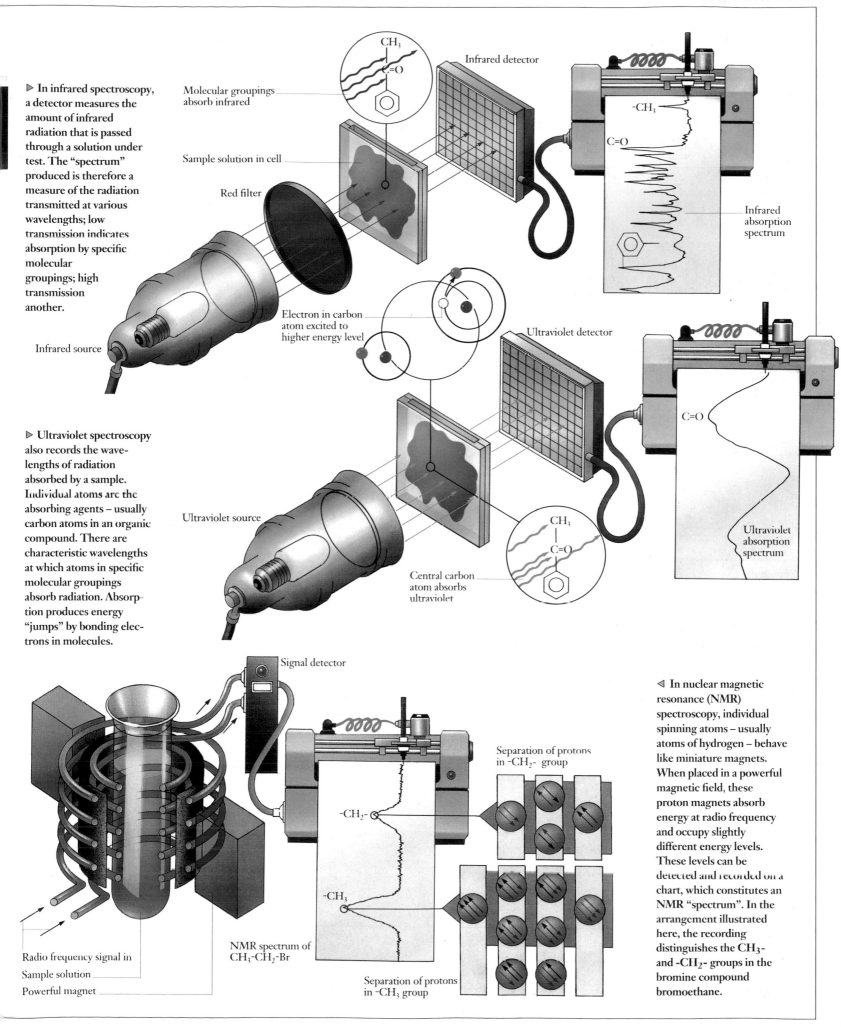

▶ In infrared spectroscopy, a detector measures the amount of infrared radiation that is passed through a solution under test. The "spectrum" produced is therefore a measure of the radiation transmitted at various wavelengths; low transmission indicates absorption by specific molecular groupings; high transmission another.

Infrared source

Molecular groupings absorb infrared

Sample solution in cell

Red filter

CH_3

$C=O$

Infrared detector

-CH_3

$C=O$

Infrared absorption spectrum

▶ Ultraviolet spectroscopy also records the wavelengths of radiation absorbed by a sample. Individual atoms are the absorbing agents – usually carbon atoms in an organic compound. There are characteristic wavelengths at which atoms in specific molecular groupings absorb radiation. Absorption produces energy "jumps" by bonding electrons in molecules.

Electron in carbon atom excited to higher energy level

Ultraviolet source

Ultraviolet detector

Central carbon atom absorbs ultraviolet

CH_3

$C=O$

$C=O$

Ultraviolet absorption spectrum

Signal detector

Radio frequency signal in

Sample solution

Powerful magnet

NMR spectrum of CH_3-CH_2-Br

-CH_2-

-CH_3

Separation of protons in -CH_2- group

Separation of protons in -CH_3 group

◀ In nuclear magnetic resonance (NMR) spectroscopy, individual spinning atoms – usually atoms of hydrogen – behave like miniature magnets. When placed in a powerful magnetic field, these proton magnets absorb energy at radio frequency and occupy slightly different energy levels. These levels can be detected and recorded on a chart, which constitutes an NMR "spectrum". In the arrangement illustrated here, the recording distinguishes the CH_3- and -CH_2- groups in the bromine compound bromoethane.

FACTFILE

PRECISE MEASUREMENT is at the heart of all science, and the several standard systems have been in use in the present century in different societies. Today, the SI system of units is universally used by scientists, but other units are used in some parts of the world. The metric system, which was developed in France in the late 18th century, is in everyday use in many countries, as well as being used by scientists; but imperial units (based on the traditional British measurement standard, also known as the foot–pound–second system), and standard units (based on commonly used American standards) are still in common use.

Whereas the basic units of length, mass and time were originally defined arbitrarily, scientists have sought to establish definitions of these which can be related to measurable physical constants; thus length is now defined in terms of the speed of light, and time in terms of the vibrations of a crystal of a particular atom. Mass, however, still eludes such definition, and is based on a piece of platinum-iridium metal kept in Sèvres, near Paris.

□ METRIC PREFIXES

Very large and very small units are often written using powers of ten; in addition the following prefixes are also used with SI units. Examples include: milligram (mg), meaning one thousandth of a gram, kilogram (kg), meaning one thousand grams.

Name	Number	Factor	Prefix	Symbol
trillionth	0.000000000001	10^{-12}	pico-	p
billionth	0.000000001	10^{-9}	nano-	n
millionth	0.000001	10^{-6}	micro-	μ
thousandth	0.001	10^{-3}	milli-	m
hundredth	0.01	10^{-2}	centi-	c
tenth	0.1	10^{-1}	deci-	d
one	1.0	10^{0}	–	–
ten	10	10^{1}	deca-	da
hundred	100	10^{2}	hecto-	h
thousand	1000	10^{3}	kilo-	k
million	1,000,000	10^{6}	mega-	M
billion	1,000,000,000	10^{9}	giga-	G
trillion	1,000,000,000,000	10^{12}	tera-	T
quadrillion	1,000,000,000,000,000	10^{15}	exa-	E

□ CONVERSION FACTORS

Conversion of METRIC units to imperial (or standard) units

To convert:	to:	multiply by:
LENGTH		
millimeters	inches	0.03937
centimeters	inches	0.3937
meters	inches	39.37
meters	feet	3.2808
meters	yards	1.0936
kilometers	miles	0.6214
AREA		
square centimeters	square inches	0.1552
square meters	square feet	10.7636
square meters	square yards	1.196
square kilometers	square miles	0.3861
square kilometers	acres	247.1
hectares	acres	2.471
VOLUME		
cubic centimeters	cubic inches	0.061
cubic meters	cubic feet	35.315
cubic meters	cubic yards	1.308
cubic kilometers	cubic miles	0.2399
CAPACITY		
milliliters	fluid ounces	0.0351
milliliters	pints	0.00176 (0.002114 for US pints)
liters	pints	1.760 (2.114 for US pints)
liters	gallons	0.2193 (0.2643 for US gallons)
WEIGHT		
grams	ounces	0.0352
grams	pounds	0.0022
kilograms	pounds	2.2046
tonnes	tons	0.9842 (1.1023 for US, or short, tons)
TEMPERATURE		
Celsius	fahrenheit	1.8, then add 32

Conversion of STANDARD (or imperial) units to metric units

To convert:	to:	multiply by:
LENGTH		
inches	millimeters	25.4
inches	centimeters	2.54
inches	meters	0.245
feet	meters	0.3048
yards	meters	0.9144
miles	kilometers	1.6094
AREA		
square inches	square centimeters	6.4516
square feet	square meters	0.0929
square yards	square meters	0.8316
square miles	square kilometers	2.5898
acres	hectares	0.4047
acres	square kilometers	0.00405
VOLUME		
cubic inches	cubic centimeters	16.3871
cubic feet	cubic meters	0.0283
cubic yards	cubic meters	0.7646
cubic miles	cubic kilometers	4.1678
CAPACITY		
fluid ounces	milliliters	28.5
pints	milliliters	568.0 (473.32 for US pints)
pints	liters	0.568 (0.4733 for US pints)
gallons	liters	4.55 (3.785 for US gallons)
WEIGHT		
ounces	grams	28.3495
pounds	grams	453.592
pounds	kilograms	0.4536
tons	tonnes	1.0161
TEMPERATURE		
fahrenheit	Celsius	subtract 32, then × 0.55556

□ SI UNITS

Now universally employed throughout the world of science and the legal standard in many countries, SI units (short for *Système International d'Unités*) were adopted by the General Conference on Weights and Measures in 1960. There are seven base units and two supplementary ones, which replaced those of the MKS (meter–kilogram–second) and CGS (centimeter–gram–second) systems that were used previously. There are also 18 derived units, and all SI units have an internationally agreed symbol.

None of the unit terms, even if named for a notable scientist, begins with a capital letter: thus, for example, the units of temperature and force are the kelvin and the newton (the abbreviations of some units are capitalized, however). Apart from the kilogram, which is an arbitrary standard based on a carefully preserved piece of metal, all the basic units are now defined in a manner that permits them to be measured conveniently in a laboratory.

Name	Symbol	Quantity	Standard
BASIC UNITS			
meter	m	length	The distance light travels in a vacuum in $\frac{1}{299,792,458}$ of a second
kilogram	kg	mass	The mass of the international prototype kilogram, a cylinder of platinum-iridium alloy, kept at Sèvres, France
second	s	time	The time taken for 9,192,631,770 resonance vibrations of an atom of cesium-133
kelvin	K	temperature	$\frac{1}{273.16}$ of the thermodynamic temperature of the triple point of water
ampere	A	electric current	The current that produces a force of 2×10^{-7} newtons per meter between two parallel conductors of infinite length and negligible cross section, placed one meter apart in a vacuum
mole	mol	amount of substance	The amount of a substance that contains as many atoms, molecules, ions or subatomic particles as 12 grams of carbon-12 has atoms
candela	cd	luminous intensity	The luminous intensity of a source that emits monochromatic light of a frequency 540×10^{-12} hertz and whose radiant intensity is $\frac{1}{683}$ watt per steradian in a given direction
SUPPLEMENTARY UNITS			
radian	rad	plane angle	The angle subtended at the center of a circle by an arc whose length is the radius of the circle
steradian	sr	solid angle	The solid angle subtended at the center of a sphere by a part of the surface whose area is equal to the square of the radius of the sphere

Name	Symbol	Quantity	Standard
DERIVED UNITS			
becquerel	Bq	radioactivity	The activity of a quantity of a radio-isotope in which 1 nucleus decays (on average) every second
coulomb	C	electric current	The quantity of electricity carried by a charge of 1 ampere flowing for 1 second
farad	F	electric capacitance	The capacitance that holds charge of 1 coulomb when it is charged by a potential difference of 1 volt
gray	Gy	absorbed dose	The dosage of ionizing radiation equal to 1 joule of energy per kilogram
henry	H	inductance	The mutual inductance in a closed circuit in which an electromotive force of 1 volt is produced by a current that varies at 1 ampere per second
hertz	Hz	frequency	The frequency of 1 cycle per second
joule	J	energy	The work done when a force of 1 newton moves its point of application 1 meter in its direction of application
lumen	lm	luminous flux	The amount of light emitted per unit solid angle by a source of 1 candela intensity
lux	lx	illuminance	The amount of light that illuminates 1 square meter with a flux of 1 lumen
newton	N	force	The force that gives a mass of 1 kilogram an acceleration of 1 meter per second per second
ohm	Ω	electric resistance	The resistance of a conductor across which a potential of 1 volt produces a current of 1 ampere
pascal	Pa	pressure	The pressure exerted when a force of 1 newton acts on an area of 1 square meter
siemens	S	electric conductance	The conductance of a material or circuit component that has a resistance of 1 ohm
sievert	Sv	dose	The radiation dosage equal to 1 joule equivalent of radiant energy per kilogram
tesla	T	magnetic flux density	The flux density (or density induction) of 1 weber of magnetic flux per square meter
volt	V	electric potential	The potential difference across a conductor in which a constant current of 1 ampere dissipates 1 watt of power
watt	W	power	The amount of power equal to a rate of energy transfer of (or rate of doing work at) 1 joule per second
weber	Wb	magnetic flux	The amount of magnetic flux that, decaying to zero in 1 second, induces an electromotive force of 1 volt in a circuit of one turn

Food additives are chemicals that are added to food to prolong its shelf life, alter its color, enhance flavor or improve nutritional value. The chemicals used can be either naturally occurring or manufactured synthetically. Most countries have laws that govern the use of food additives. In the United States, the Food and Drug Administration is responsible for enforcing these laws and, if necessary, bans the use of specific chemicals. For example, in 1970 the FDA banned artifical sweeteners called cyclamates because they had been shown to be cancer-causing agents in laboratory animals. In Europe, the European Union (EU), assess all known additives. The EU grants a number, known as an E-number, to those approved as safe. However, it should be noted that some people are allergic to permitted additives, and that the safety of some of the listed substances is not accepted in all countries. To help consumers recognize which chemicals have been added to the products they buy, the numbers must appear on the food label.

There are five major classes of food additives. Antioxidants, as their name implies, prevent fats and oils from oxidizing and therefore going rancid. They also help to prevent the discoloration of foods.

Coloring agents improve the colors of processed foods and even synthetic ones (such as margarine). Canned peas, for example, lose their natural color and manufacturers may add a coloring agent to restore their bright green appearance. Other colors are applied to the skins of fruits to enhance their appearance of ripeness.

Emulsifiers and stabilizers prevent the components of mixed foods from separating, particularly with emulsions such as ice cream and mayonnaise. An example is pectin, added to jams and jellies to thicken them.

Preservatives prevent the growth of organisms, such as bacteria, that might spoil food products. Many preservatives are similar to antioxidants. Others include traditional substances such as salt and saltpeter.

Sweeteners and flavor enhancers bring out the taste of foods although, like monosodium glutamate (MSG), they may have no flavor of their own.

In addition, nutritional supplements may be added to manufactured products to make them more nourishing. Vitamins, for example, are usually added to flour (and therefore to products made from it), margarine and some milk products. Trace minerals may also be added.

Name and number		Source
ANTIOXIDANTS		
synthetic alpha-tocopherol	E107	Prepared synthetically in the laboratory
sulfur dioxide	E220	Occurs naturally, also produced chemically by the combustion of sulfur or gypsum
L-ascorbic acid (vitamin C)	E301	Occurs naturally in many fresh fruits and vegetables, also manufactured by biological synthesis involving hydrogenation or fermentation
sodium L-ascorbate	E301	Prepared synthetically from ascorbic acid (E300)
extracts of natural origin rich in tocopherols (vitamin E)	E306	Extracted from soya bean oil, wheat germ, rice germ, cottonseed, maize (corn) and green leaves
propyl gallate	E310	Produced from tannins extracted from nut galls or by hydrolysis of the enzyme tannase, which occurs in spent fungal broths of *Aspergillus niger* and *Penicillium glaucum*
octyl gallate	E311	Obtained by acid or alkaline hydrolysis of the tannins extracted from nut galls or by hydrolysis of the enzyme tannase.
dodecyl gallate	E312	Produced from tannins or the enzyme tannase
butylated hydroxyanisole (BHA)	E320	Prepared from *p*-methoxyphenol and isobutene
butylated hydroxytoluene (BHT)	E320	Prepared synthetically from *p*-cresol and isobutylene
lecithin	E322	Present in all living cells. Sources for food use include egg yolk and leguminous seeds (including peanuts) and maize (corn)
sodium lactate	E325	Produced from lactic acid (E270)

Name and number		Source
COLORING AGENTS		
tartrazine	E102	A synthetically produced yellow azo dye
quinoline yellow	E104	A synthetically produced greenish-yellow dye
sunset yellow	E110	A synthetically produced yellow azo dye
cochineal	E120	A red dye produced from the bodies of scale insects (*Dactilopius coccus*)
carmoisine	E122	A synthetically produced purple-red azo dye
amaranth	E123	A synthetically produced puple-red azo dye
ponceau 4R	E 124	A synthetically produced red azo dye
erythrosine	E127	A synthetically produced cherry pink to red dye
chlorophyll	E140	A green pigment extracted from leaves
copper chlorophyll	E141	A copper complex prepared in the laboratory from chlorophyll (E140)
caramel	E150	Prepared by heating sugar syrups in the presence of food-grade ammonia, ammonium sulfate, sulfur dioxide and/or sodium hydroxide
annatto; bixin; Norbixin	E160(b)	A yellow to peach or red dye produced from the seed coat of the annatto tree (*Bixa orellana*)
betanin	E162	A red dye obtained from beetroot, also known as beetroot red

Name and number		Source
EMULSIFIERS AND STABILIZERS		
alginic acid	E400	Extracted from brown seaweeds, especially species of *Laminaria*
propane-1,2-diol alginate (propylene glycol alginate)	E405	Derived from brown seaweeds
agar	E406	Derived from red seaweeds
carrageenan	E407	Derived from species of red seaweeds
carob gum	E410	Extracted from seed pods (beans) of the honey locust tree
gum arabic (acacia)	E414	Derived from the stems and branches of the *Acacia senegal* tree and related species
pectin	E440(a)	Derived commercially from apple residues from cider making, and from orange pith
microcrystalline cellulose; alpha cellulose	E460	Produced by chemical fragmentation of the cellulose walls of plant fibers into microscopic crystals
mono- and di-glycerides of fatty acids	E471	Prepared commercially from glycerin and fatty acids
polyglycerol esters of fattyacids	E475	Prepared in the laboratory
polyglycerol esters of polycondensed fatty acids of castor oil	E476	Prepared from castor oil and glycerol esters
sodium stearoyl-1, 2-lactylate	E481	Prepared in the laboratory from various vegetable oils
PRESERVATIVES		
sorbic acid	E200	A natural acid extracted from the berries of the mountain ash (rowan)
benzoic acid	E210	Naturally occurring, but prepared commercially by chemical synthesis
sodium benzoate	E211	The sodium salt of benzoic acid
calcium benzoate	E213	The calcium salt of benzoic acid
sulfur dioxide	E220	Occurs naturally and produced commercially by chemical synthesis
sodium sulfite	E221	A sodium salt of sulfurous acid
hexamine	E239	Hexamethylenetetramine, made in the laboratory from methanal (formaldehyde) and ammmonia
potassium nitrite	E249	Potassium salt of nitrous acid

Name and number		Source
sodium nitrite	E250	Derived from sodium nitrate by chemical or bacterial means
sodium nitrate	E251	A naturally occurring mineral
potassium nitrate	E252	A naturally occurring mineral, also manufactured from waste animal and vegetable matter
acetic acid	E260	Manufactured chemically using methanol and carbon monoxide, or from ethanol by oxidation. Also produced by the action of the bacterium *Acetobacter*
SWEETENERS AND FLAVOR ENHANCERS		
sorbitol	E420	Naturally occurring, but produced commercially from glucose by high-pressure hydrogenation or electrolytic reduction
mannitol	E421	Prepared from seaweed
L-glutamic acid	E620	Naturally occurring, but prepared commercially by the fermentation of a carbohydrate solution by a bacterium
monosodium glutamate (MSG)	E621	Naturally occurring, but prepared commercially by fermentation using molasses from cane or beet sugar
monopotassium glutamate	E622	Prepared synthetically
calcium dihydrogen di-L-glutamate	E623	Prepared synthetically
sodium guanylate	E627	Naturally occurring, but prepared synthetically for commercial use
sodium 5'-inosinate	E631	Prepared from meat extract and dried sardines
sodium 5'-ribonucleotide	E635	A mixture of di-sodium guanylate (E627) and sodium 5'-inosinate (E631)
maltol	E636	A naturally occurring substance which is also obtained chemically by alkaline hydrolysis of streptomycin salt
sodium saccharin		Prepared in the laboratory by oxidizing a derivative of toluene
aspartame		Prepared in the laboratory from aspartic acid, a naturally occurring amino acid in proteins

Some of the elements have been known since prehistoric times, even if the people at that time did not have a concept of what an element is. They included carbon, copper, gold, iron, lead, mercury, silver, sulfur and tin. Their names derive from Latin *carbos* (= charcoal), symbol C; Latin *Cypros* (= Cyprus), symbol Cu; Anglo-Saxon *gold* (Latin *aurum*), symbol Au; Anglo-Saxon *iren* (Latin *ferrum*), symbol Fe; Anglo-Saxon *lead* (Latin *plumbum*), symbol Pb; Mercury in mythology (Latin *hydragyrum*), symbol Hg; Anglo-Saxon *seolfor* (Latin *argentum*), symbol Ag; Latin *sulfur*, symbol S; and Anglo-Saxon *tin* (Latin *stanum*), symbol Sn. Zinc (German *Zink*, symbol Zn) has been used from before 1500, but its discoverer is unknown.

Date	Name and symbol	Discovered by	Meaning of name *
1450	antimony Sb	Valentine	Lat *antimonium*, formerly *stibium*
	bismuth Bi	Valentine	Ger *Wismut*
1649	arsenic As	Schröder	Lat *arsenicum*
1669	phosphorus P	Henning Brand	Lat ex Gk = light-bearing
1735	cobalt Co	Georg Brandt	Ger *Kobold* = (mine) goblin
1741	platinum Pt	William Wood	Sp *platina* = silver
1751	nickel Ni	Axel Cronstedt	Swe (kuppar) – nickel
1766	hydrogen H	Henry Cavendish	Gk *hydor* = water + gen
1771	fluorine F	Carl Scheele	Lat *fluo* = flow
1772	nitrogen N	Daniel Rutherford	Gk *nitron* = saltpeter
1774	manganese Mn	Johan Gahn	Lat *magnes* = magnet
	oxygen O	John Priestley	Gk *oxys* = acid + gen
	chlorine Cl	Carl Scheele	Gk *chloros* = green
1778	molybdenum Mo	Peter Hjelm	Gk *molybdos* = lead
1782	tellurium Te	Franz von Reichenstein	Lat *tellus* = earth
1783	tungsten W	Don Fausto and Don Juan d'Elhuyar	Swe heavy stone (formerly wolfram)
1789	titanium Ti	William Gregor	Titanes (Lat myth)
	uranium U	Martin Klaproth	Uranus (planet)
1794	yttrium Y	Johan Gadolin	Ytterby (town in Sweden)
1797	chromium Cr	Louis Vauquelin	Gk *chromos* = color
1801	niobium Nb	Charles Hatchett	Niobe (Gk myth)
1802	tantalum Ta	Anders Ekeberg	Tantalus (Gk myth)
1803	cerium Ce	Jöns Berzelius and Wilhelm Hisinger	Ceres (asteroid)
1803	iridium Ir	Smithson Tennant	Lat *iris* = rainbow
	osmium Os	Smithson Tennant	Gk *osme* = odor
1804	palladium Pd	William Wollaston	Pallas (asteroid)
	rhodium Rh	William Wollaston	Gk *rhodon* = rose
1807	potassium K	Humphry Davy	Eng potash = Lat *kalium*
	sodium Na	Humphry Davy	Eng soda = Lat *natrium*
1808	barium Ba	Humphry Davy	Gk *barys* = heavy
	boron B	Humphry Davy	BORax+carbON
	calcium Ca	Humphry Davy	Lat *calx* = lime
	strontium Sr	Humphry Davy	Strontian (place in Scotland)
1811	iodine I	Bernard Courtois	Gk *iodos* = violet
1817	cadmium Cd	Friedrich Strohmeyer	Gk *kadmeia* = calamine
	lithium Li	Johan Arfvedson	Gk *lithos* = stone
	selenium Se	Jöns Berzelius	Gk *selene* = Moon
1824	silicon Si	Jöns Berzelius	Lat *silex* = flint
1824	zirconium Zr	Jöns Berzelius	P *zargun* = gold-colored
1825	bromine Br	Antoine Balard	Gk *bromos* = stench
1827	aluminum Al	Friedrich Wöhler	Lat *alumen* = alum
1828	beryllium Be	Friedrich Wöhler	Gk *beryllion* = beryl (mineral)
	thorium Th	Jöns Berzelius	Thor (Norse myth)
1829	magnesium Mg	Antoine Bussy	Magnesia (in Italy)
1830	vanadium V	Nils Sefström	Vanadis (Norse myth)
1839	lanthanum La	Carl Mosander	Gk *lanthano* = conceal
1843	erbium Er	Carl Mosander	Ytterby (town in Sweden)
	terbium Tb	Carl Mosander	Ytterby (town in Sweden)
1845	ruthenium Ru	Carl Claus	Lat *Ruthenia* = Russia

□ COMMON ELEMENTS ON EARTH AND IN THE BODY

Ninety-two different elements occur naturally on Earth. The commonest is oxygen, which makes up 46.6 percent by weight of the Earth's crust and 20.95 percent by volume of the atmosphere. Also the oceans consist mainly of water, a compound that is 88.89 percent oxygen. Oxygen is also the commonest element in the human body – making up 65 percent of our weight. Carbon and hydrogen are the next most abundant.

The table lists the abundances of 16 elements in the Earth's crust and their occurrence in the atmosphere, the sea and the human body. Figures are in percentages except for the sea's composition, which is of the dissolved substances expressed in parts per million (ppm).

Name and symbol	Atomic number	% of Earth's crust	% Earth's atmosphere	ppm of the sea	% of the human body	Function in the body
Hydrogen H	1	0.14	Trace	–	9.5	Electron carrier, component of water
Carbon C	6	0.03	0.03	–	18.5	Backbone of organic molecules
Nitrogen N	7	Trace	78.1	–	3.3	Component of all nucleic acids and proteins
Oxygen O	8	46.6	20.95	–	65.0	Required for cellular respiration, component of water
Fluorine F	9	0.07	–	0.003	Trace	
Sodium Na	11	2.8	–	10,556	0.2	Important in nerve functioning
Magnesium Mg	12	2.1	–	1272	0.1	Component of many energy-transferring enzymes
Aluminum Al	13	8.1	–	–	Trace	
Silicon Si	14	27.7	–	–	Trace	
Phosphorus P	15	0.07	–	–	1.0	Backbone of nucleic acids; component of bones and teeth
Sulfur S	16	0.03	–	7.7	0.3	Component of some proteins
Chlorine Cl	17	0.01	–	18,980	0.2	Negative ion bathing cells
Potassium K	19	2.6	–	1.1	0.4	Important in nerve functioning
Calcium Ca	20	3.6	–	400	1.5	Component of bones and teeth
Manganese Mn	25	0.1	–	–	Trace	
Iron Fe	26	5.0	–	–	Trace	Component of hemoglobin in blood

1860	cesium Ca	Robert Bunsen	Lat *caesium* = blue-gray
1861	rubidium Rb	Robert Bunsen and Gustav Kirchhoff	Lat *rubidus* = red
1861	thallium Tl	William Crookes	Gk *thallos* = budding twig
1863	indium In	Ferdinand Reich and Hyeronimus Richter	Indigo
1875	gallium Ga	Paul de Boisbauderan	Lat *Gallia* = France
1878	ytterbium Yb	Jean Marignac	Ytterby (town in Sweden)
1879	holmium Ho	Per Cleve	Lat *Holmia* = Stockholm
	scandium Sc	Lars Nilson	Scandinavia
	samarium Sm	Paul de Boisbauderan	Samarski (Russian mineralogist)
	thulium Tm	Per Cleve	Swe *Thule* = northland
1885	gadolinium Gd	Jean Marignac	Gadolin (Fin chemist)
	neodymium Nd	Carl Welsbach	Gk *neos* = new + *didymos* = twin
	praseodymium Pr	Carl Welsbach	Gk *prasios* = green + *didymos* = twin
1886	dysprosium Dy	Paul de Boisbauderan	Gk *dysprositos* = difficult to approach
	germanium Ge	Clemens Winkler	Lat *Germania* = Germany
1894	argon Ar	Lord Rayleigh and William Ramsay	Gk *argos* = inactive
1895	helium He	William Ramsay and William Crookes	Gk *helios* = Sun
1896	europium Eu	Eugène Demarçay	Europe
1898	krypton Kr	William Ramsay and Morris Travers	Gk *kryptos* = hidden
	neon Ne	William Ramsay and Morris Travers	Gk *neos* = new
	xenon Xe	William Ramsay and Morris Travers	Gk *xenos* – stranger
	polonium Po	Marie Curie	Poland
	radium Ra	Marie Curie	Lat *radius* = ray

1899	actinium Ac	André Debierne	Gr *aktis* = ray
1900	radon Rn	Ernst Dorn	RADium emanatiON
1907	lutetium Lu	Georges Urbain and Carl Welsbach	Lat *Lutetia* = Paris
1917	protactinium Pa	Lise Meitner and Otto Hahn	Gk *protos* = first + actinium
1923	hafnium Hf	Dirk Coster and György Heversey	Lat *Hafnia* – Copenhagen
1925	rhenium Re	Walter Noddack and Ida Tacke	Ger *Rhein* = Rhine
1937	technetium Tc	Perrier and Emilio Segrè	Gk *technetos* = artificial
1939	francium Fr	Marguerite Perey	France
1940	astatine At	Emilio Segrè *et al*	Gk *astatos* = unstable
	neptunium Np	Philip Abelson and Edwin McMillan	Neptune (planet)
	promethium Pm	J. Marinsky *et al*	Prometheus (Gk myth)
1941	plutonium Pu	Glen Seaborg and Edwin Mattison	Pluto (planet)
1944	americium Am	Glen Seaborg *et al*	America
	curium Cm	Glen Seaborg *et al*	Mme Curie
1950	berkelium Bk	Glen Seaborg *et al*	Berkeley, Cal.
	californium Cf	Glen Seaborg *et al*	California
1952	einsteinium Es	Glen Seaborg *et al*	Einstein
	fermium Fm	Glen Seaborg *et al*	Fermi
1955	mendelevium Md	Glen Seaborg, Albert Ghiorso *et al*	Mendeleyev
1957	nobelium No	Glen Seaborg, Albert Ghiorso *et al*	Nobel
1961	lawrencium Lr	Albert Ghiorso *et al*	Lawrence
1969	rutherfordium Rf	University of California	Rutherford
1970	hahnium Ha	University of California	Hahn

* AS = Anglo-Saxon; Eng = English; Fin = Finnish; Ger = German; Gk = Greek; Lat = Latin; P = Persian; Sp = Spanish; Swe = Swedish

☐ COMMON CYCLIC COMPOUNDS

Many organic compounds have molecules that contain rings of atoms – the so-called cyclic compounds. If all the atoms in the ring are the same, as in cyclohexane (C_6H_{12}) and benzene (C_6H_6), they are known as homocyclic. If there are two or more different atoms in the ring, the compound is heterocyclic. The formulas of some heterocyclic compounds are shown here. Those containing nitrogen atoms occur in important biological compounds. For example, purine forms the skeleton of two of the bases of DNA, the fundamental substance that carries the genetic codes for all living organisms.

Simple heterocyclic compounds

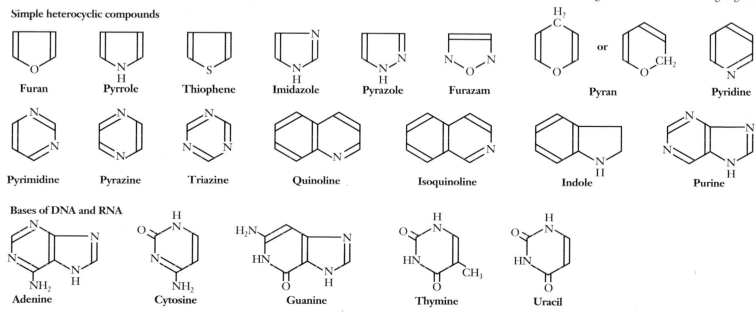

Furan   Pyrrole   Thiophene   Imidazole   Pyrazole   Furazam   Pyran   Pyridine

Pyrimidine   Pyrazine   Triazine   Quinoline   Isoquinoline   Indole   Purine

Bases of DNA and RNA

Adenine   Cytosine   Guanine   Thymine   Uracil

The chemical elements in the Periodic Table are arranged in order of increasing atomic number. In this version of the Table, the atomic number is shown in the top right-hand corner of each box. The numbers down the left-hand side of each box are the numbers of electrons in each "shell", or set of orbitals, surrounding the atom's nucleus.

The vertical columns of the Table consist of groups, while the horizontal rows are known as periods. Elements occupying the same group have similar chemical properties, with the more reactive members near the top. There is also a gradation of properties across periods, from the highly metallic elements on the left of the Table to the non-metals on the right. Again, neighboring elements share some properties.

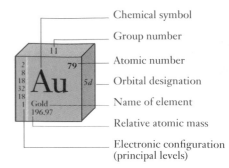

Chemical symbol
Group number
Atomic number
Orbital designation
Name of element
Relative atomic mass
Electronic configuration (principal levels)

s block

I	II
1	2

2,1	Li 3		2,2	Be 4
	Lithium 6.94			Beryllium 9.01
2,8,1	Na 11		2,8,2	Mg 12
	Sodium 22.99			Magnesium 24.31

d block   Transition elements

3	4	5	6	7	8	9	10	11	12

2,8,8,1 K 19 Potassium 39.10	2,8,8,2 Ca 20 Calcium 40.08	2,8,9,2 Sc 21 Scandium 44.96	2,8,10,2 Ti 22 Titanium 47.90	2,8,11,2 V 23 Vanadium 50.94	2,8,13,1 Cr 24 Chromium 51.99	2,8,13,2 Mn 25 Manganese 54.94	2,8,14,2 Fe 26 Iron 56.85	2,8,15,2 Co 27 Cobalt 58.93	2,8,16,2 Ni 28 Nickel 58.71	2,8,18,1 Cu 29 Copper 63.55	2,8,18,2 Zn 30 Zinc 65.37

3d

| 2,8,18,8,1 Rb 37 Rubidium 85.47 | 2,8,18,8,2 Sr 38 Strontium 87.62 | 2,8,18,9,2 Y 39 Yttrium 88.90 | 2,8,18,10,2 Zr 40 Zirconium 91.22 | 2,8,18,12,1 Nb 41 Niobium 92.91 | 2,8,18,13,1 Mo 42 Molybdenum 96.94 | 2,8,18,13,2 Tc 43 Technetium (98) | 2,8,18,15,1 Ru 44 Ruthenium 101.07 | 2,8,18,16,1 Rh 45 Rhodium 102.90 | 2,8,18,18 Pd 46 Palladium 106.4 | 2,8,18,18,1 Ag 47 Silver 107.87 | 2,8,18,18,2 Cd 48 Cadmium 112.40 |

4d

| 2,8,18,18,8,1 Cs 55 Cesium 132.90 | 2,8,18,18,8,2 Ba 56 Barium 137.34 | 2,8,18,32,9,2 Lu 71 Lutetium 174.97 | 2,8,18,32,10,2 Hf 72 Hafnium 178.49 | 2,8,18,32,11,2 Ta 73 Tantalum 180.95 | 2,8,18,32,12,2 W 74 Tungsten 180.95 | 2,8,18,32,13,2 Re 75 Rhenium 186.2 | 2,8,18,32,14,2 Os 76 Osmium 190.2 | 2,8,18,32,15,2 Ir 77 Iridium 192.2 | 2,8,18,32,17,1 Pt 78 Platinum 195.09 | 2,8,18,32,18 Au 79 Gold 196.97 | 2,8,18,32,18,2 Hg 80 Mercury 200.59 |

5d

| 2,8,18,32,18,8,1 Fr 87 Francium (223) | 2,8,18,32,18,8,2 Ra 88 Radium (226) | 2,8,18,32,9,2 Lr 103 Lawrencium (257) | 2,8,18,32,10,2 Rf 104 Rutherfordium (261) | 2,8,18,32,11,2 Ha 105 Hahnium (262) |

6d

f block   Inner transition elements (lanthanides and actinides)

| 2,8,18,18,9,2 La 57 Lanthanum 138.91 | 2,8,18,19,9,2 Ce 58 Cerium 140.12 | 2,8,18,21,8,2 Pr 59 Praseodymium 140.9 | 2,8,18,22,8,2 Nd 60 Neodymium 144.24 | 2,8,18,23,8,2 Pm 61 Promethium (147) | 2,8,18,24,8,2 Sm 62 Samarium 150.35 | 2,8,18,25,8,2 Eu 63 Europium 151.96 | 2,8,18,25,9,2 Gd 64 Gadolinium 157.25 | 2,8,18,27,8,2 Tb 65 Terbium 158.92 | 2,8,18,28,8,2 Dy 66 Dysprosium 162.50 |

| 2,8,18,32,18,9,2 Ac 89 Actinium (227) | 2,8,18,32,18,10,2 Th 90 Thorium 232.04 | 2,8,18,32,21,9,2 Pa 91 Protactinium (231) | 2,8,18,32,21,9,1 U 92 Uranium 238.03 | 2,8,18,32,23,8,2 Np 93 Neptunium (237) | 2,8,18,32,24,8,2 Pu 94 Plutonium (242) | 2,8,18,32,25,9,2 Am 95 Americium (243) | 2,8,18,32,25,9,2 Cm 96 Curium (247) | 2,8,18,32,27,8,2 Bk 97 Berkelium (247) | 2,8,18,32,28,8,2 Cf 98 Californium (249) |

Electrons surrounding an atom's nucleus occupy regions of space called atomic orbitals, which have the characteristic shapes shown here. The shapes are obtained by plotting in three dimensions the mathematical function that represents the probability that an electron of a particular energy level will be found at a particular location. For example, s-orbitals take the shape of a spherical shell symmetrically surrounding the nucleus. Each s-orbital can hold one or two electrons. There are three double-lobed p-orbitals, each again holding one or two electrons (giving a maximum of six in any of the p energy levels). Most of the five d-orbitals have four lobes, but they can still hold only up to two electrons each (making a possible total of 10 in the d levels). Finally there are seven different f-orbitals, with a total capacity of 14 electrons.

When atoms come together to form a covalent bond, the atomic orbitals combine to form molecular orbitals. These vary greatly in shape from compound to compound. For example, a hydrogen atom has a single electron in a spherical 1s orbital. When two hydrogen atoms combine, their singly occupied orbitals merge to form an egg-shaped sigma orbital. Carbon, on the other hand, has four electrons in its outer (bonding) orbitals: one in a 2s orbital and one in each of the 2p orbitals. These combine to form four symmetrical hybrid orbitals designated sp^3, which are arranged in space pointed toward the corners of a regular tetrahedron. Thus when a carbon atom combines with four hydrogen atoms to form a molecule of methane (CH_4), the resulting compound has a tetrahedral shape. In ethene (ethylene, C_2H_4), a different kind of hybridization occurs. This time, the carbon's 2s and two of its 2p orbitals hybridize to form three sp^2 orbitals, which point toward the corners of of an equilateral triangle. Two of these (one from each carbon atom) overlap with each other to form a carbon–carbon bond, and the other four combine with the 2s orbital of the four hydrogen atoms. The remaining 2p atomic orbitals – one on each carbon – also combine to form a pair of pi orbitals, so completing the double C=C bond in the molecule.

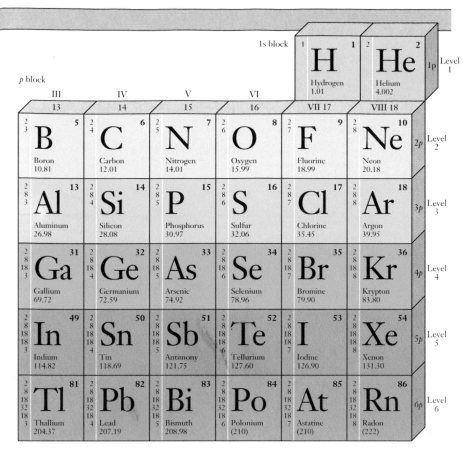

1s block

p block

| III | IV | V | VI | VII 17 | VIII 18 |

(Periodic table p-block: B, C, N, O, F, Ne; Al, Si, P, S, Cl, Ar; Ga, Ge, As, Se, Br, Kr; In, Sn, Sb, Te, I, Xe; Tl, Pb, Bi, Po, At, Rn; with H and He in the 1s block)

(Lanthanide/actinide block: Ho 67, Er 68, Tm 69, Yb 70; Es 99, Fm 100, Md 101, No 102)

☐ THE AUFBAU PRINCIPLE

Electrons surrounding an atom's nucleus occupy shells, each made up of one or more orbitals. Each orbital has a characteristic energy level, and the diagram BELOW shows the order of energies, and thus the order in which electrons theoretically fill up the shells. The first shell – with a single 1s orbital – can hold a maximum of only two electrons, and it is filled by the time we reach helium (He). The next two shells hold up to eight electrons (two in the 2s and 3s orbitals, and six in the 2p and 3p levels), and these shells are filled at neon (Ne) and argon (Ar). The fourth and fifth shells hold a maximum of 18 electrons (up to 10 going in the 3d and 4d orbitals), and are complete at krypton (Kr) and xenon (Xe). Finally the sixth and seventh shells can accommodate up to 32 electrons (with as many as 14 in the 4f and 5f orbitals); the fifth is complete at radon (Rn), but it is unlikely that scientists will ever complete the 6d shell through as far as element 118.

Orbital

(Orbital energy-level filling diagram, from top to bottom: 7p, 6d, 5f, 7s, 6p, 5d, 4f, 6s, 5p, 4d, 5s, 4p, 3d, 4s, 3p, 3s, 2p, 2s, 1s)

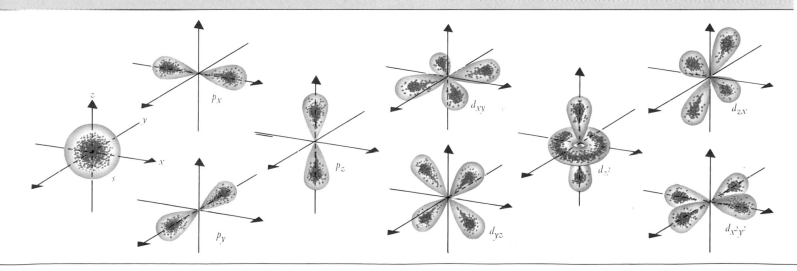

(Orbital shapes: s, p_x, p_y, p_z, d_{xy}, d_{yz}, d_{zx}, d_{z^2}, d_{x^2-y^2})

□ COMMON CHEMICALS

Originally, chemical substances were given familiar names that usually reflected their appearance or properties – some date back to the days of alchemy. For modern scientific use, these names were replaced by systematic ones, recognized internationally. But many of the common names are still used by non-scientists and in old-fashioned formulations, and the following table lists some of these with their systematic names, formulas and everyday applications. Some of the substances can be found in the home or garage, but it should be remembered that many are dangerously corrosive or poisonous.

Common name	Chemical name	Formula	Some common uses
Alum	Aluminum potassium sulfate	$Al_2(SO4)_3.K_2SO_4.24H_2O$	Dyeing, tanning, leather finishing, paper sizing, styptic
Aqua fortis	Nitric acid	HNO_3	Metal cleaning, manufacture of fertilizers and explosives
Aqua regia	Hydrochloric and nitric acid	$HCl + HNO_3$	Solvent for gold and platinum
Baking soda	Sodium hydrogen carbonate	$NaHCO_3$	Antacid, cooking, fire extinguishers, paper making, tanning
Blackboard chalk	Calcium sulfate	$CaSO_4.2H_2O$	Chalk, manufacture of ceramics, paint, sulfuric acid
Black lead	Graphite	C	Metal polish, lubricant, pencils, electrical contacts
Blue vitriol	Copper(II) sulfate	$CuSO_4.5H_2O$	Fungicide, timber preservation, electroplating, dyeing
Borax	Disodium tetraborate	$Na_2B_4O_7.10H_2O$	Ant killer, flux, glass-making, antiseptic
Butter of zinc	Zinc chloride	$ZnCl$	Solder flux, dehydrating agent, catalyst, dry batteries
Calamine	Zinc carbonate	$ZnCO_3$	Powder or lotion to treat sunburn, skin disorders
Calomel	Mercury(I) chloride	Hg_2Cl_2	Fungicide, insecticide, purgative, batteries
Carbolic acid	Phenol	C_6H_5OH	Disinfectant, manufacture of plastics, dyes, explosives
Caustic potash	Potassium hydroxide	KOH	Soap-making, batteries, absorbing acid gases
Caustic soda	Sodium hydroxide	$NaOH$	Drain cleaner, making soap, paper, aluminum, petrochemicals
Chile saltpeter	Sodium nitrate	$NaNO_3$	Fertilizer, making nitric acid
Chinese white	Zinc oxide	ZnO	Pigment, skin ointment, making ceramics, plastics
Common salt	Sodium chloride	$NaCl$	Seasoning, food preservation, making sodium carbonate and hydroxide
Corrosive sublimate	Mercury(II) chloride	$HgCl_2$	Pesticide, making other mercury compounds
Cream of tartar	Potassium hydrogen tartrate	$HOOC(CHOH)_2COOK$	Baking powder, acidity-controlling food additive
Epsom salt(s)	Magnesium sulfate	$MgSO_4.7H_2O$	Laxative, fireproofing fabrics, tanning, making fertilizers, matches
Formalin	Methanal solution	$HCHO$	Preservative, germicide, sterilizing agent
Glauber's salts	Sodium sulfate	$Na_2SO_4.10H_2O$	Laxative, making glass, wood pulp, dyeing
Glycol	Ethane-1,2-diol	CH_2OHCH_2OH	Antifreezes, coolants, making polyester fibers and plasticizers
Green vitriol	Iron(II) sulfate	$FeSO_4$	Anemia treatment, tanning, making inks
Hypo	Sodium thiosulfate	$Na_2S_2O_3$	Fixing in photography, dyeing, preventing fermentation
Jeweler's rouge	Iron(III) oxide	Fe_2O_3	Abrasive, pigment
Killed spirits of salt	Zinc chloride	$ZnCl$	Solder flux, dehydrating agent, catalyst, dry batteries
Lamp black	Carbon	C	Pigment, filler for rubber in vehicle tires
Lime	Calcium oxide	CaO	Gardening, making glass, paper, steel
Litharge	Lead(II) oxide	PbO	Paint drier, pigment, making ceramics, car batteries
Metaldehyde	Ethanal tetramer	$[CH_3CHO]_4$	Camping fuel, slug killer
Milk of magnesia	Magnesium hydroxide	$Mg(OH)_2$	Antacid, refractory, heat insulation, reflective coatings
Mineral chalk	Calcium carbonate	$CaCO_3$	Making lime and sodium carbonate
Oil of vitriol	Sulfuric acid	H_2SO_4	Battery acid, pickling steel, making fertilizers, detergents
Plaster of Paris	Calcium sulfate	$CaSO_4.\frac{1}{2}H_2O$	Casts and molds, pottery, builder's plaster
Plumbago	Graphite	C	Metal polish, lubricant, pencils, electrical contacts
Potash	Potassium carbonate	KCO_3	Fertilizer, dyeing, wood finishing, glass, soap
Prussian blue	Iron(III) hexacyanoferrate(II)	$Fe_4[Fe(CN)_6]_3$	Pigment
Prussic acid	Hydrocyanic acid	HCN	Poison, fumigant, electroplating, making synthetic fibers
Red lead	Dilead(II), lead(IV) oxide	Pb_3O_4	Pigment in corrosion-resistant paints, glass making
Rochelle salt	Potassium sodium tartrate	$KOOC(CH)H_2COONa$	Microphones and battery-free electric cigarette lighters
Sal ammoniac	Ammonium chloride	NH_4Cl	Flashlight batteries, treatment of textile fibers, galvanizing flux
Saltpeter	Potassium nitrate	KNO_3	Gunpowder, food preservative, fertilizers
Sal volatile	Ammonium carbonate	$(NH_4)_2CO_3$	Smelling salts, baking powders, wool finishing, expectorant
Slaked lime	Calcium hydroxide	$Ca(OH)_2$	Gardening, water treatment, making mortar, bleaching powder
Soda ash	Anhydrous sodium carbonate	Na_2CO_3	Glass-making, food additive, photography, textile treatment
Soda water	Carbonic acid	H_2CO_3	Soft drink
Spirits of salt	Hydrochloric acid	HCl	Metal cleaning, industrial food preparation, making PVC
Sugar of lead	Lead(II) ethanoate	$(CH_3COO)_2Pb$	Dyeing, oxidizing agent
Superphosphate	Monocalcium phosphate	$Ca(H_2PO_4)_2$	Fertilizer
Vinegar	Dil. ethanoic acid	CH_3COOH	Flavoring, pickling agent
Washing soda	Sodium carbonate	Na_2CO_3	Cleanser, glass-making, photography, textile treatment
Water glass	Sodium silicate	$Na_2SiO_4.xH_2O$	Egg preservative, sizing, making silica gel (drying agent)
White lead	Lead(II) carbonate hydroxide	$2PbCO_3.Pb(OH)_2$	Pigment

☐ COMMON ALLOYS

Name	Typical composition (may vary)	Uses
Babbit metal	70% tin, 20% lead, 7% antimony, 3% copper	Bearings
Brass	70% copper, 30% zinc	Door furniture, musical instruments
Brazing solder	45% copper, 35% zinc, 30% silver	Joining metals
Britannia metal	85% tin, 15% antimony	Bearings, decorative metalware
Bronze	95% copper, 5% tin	Statuary, bells
Cast iron	95% iron, 3% carbon	Machine parts, castings
Cobalt steel	70% iron, 17% tungsten, 10% cobalt, 3% chromium	High-speed tools
Coinage bronze	95% copper, 4% tin, 1% zinc	Coins
Coinage gold	90% gold, 10% copper	Coins
Coinage silver	90% silver, 10% copper	Coins
Cupronickel	74% copper, 25% nickel	Coins
Dental amalgam	70% mercury, 30% copper	Tooth fillings
Dental gold	60% gold, 25% copper, 15% silver	Tooth fillings
Duralumin	96% aluminum, 4% copper	Aerospace parts
Electrum	70% gold, 30% silver	Jewelry
Elinvar	52% iron, 36% nickel, 12% chromium	Watch springs
German silver	55% copper, 25% zinc, 29% nickel	Jewelry, silver-plated metalware
Gun metal	90% copper, 8% tin, 2% zinc	Bearings
Invar	63.8% iron, 36% nickel, 0.2% carbon	Springs for clocks and watches

Name	Typical composition (may vary)	Uses
Pewter	80% tin, 20% lead	Decorative metalware
Manganin	82% copper, 16% magnanese, 2% nickel	Electric heater elements
Mild steel	99.96% iron, 0.4% carbon	General steelwork
Monel metal	65% nickel, 35% copper	Acid-resistant vessels
Mumetal	78% nickel, 17% iron, 5% copper	Transformer cores
Muntz metal	60% copper, 39% zinc, 1% lead	Forgings, large bolts
Nichrome	80% nickel, 20% chromium	Electric heater elements
Nickel silver	*See* German silver	
Permalloy	78.5% nickel, 21.5% iron	Transformer cores
Phosphor bronze	90% copper, 9% tin, 1% phosphorus	Gear wheels, springs
Pinchbeck	90% copper, 10% zinc	Decorative metalwork
Solder	70% tin, 30% lead	Joining metals, wiring circuits
Speculum	66% copper, 34% tin	Scientific instruments
Stainless steel	70% iron, 20% chromium, 10% nickel	Cutlery, car parts
Tool steel	90% iron, 7% molybdenum, 3% chromium	Hammers, chisels, drills
Type metal	75% lead, 15% antimony, 10% tin	Type for printing
White metal	60% tin, 25% lead, 10% copper, 5% antimony	Bearings, small castings
Wood's metal	50% bismuth, 25% lead, 12.5% cadmium, 12.5% tin	Fusible heads in sprinkler systems

☐ FURTHER READING

Asimov, I. *Asimov on Chemistry* (Macdonald and Jones, London, 1975)

Atkins, P.W. *Atoms, Electrons and Change* (Scientific American Library, New York, 1991)

Atkins, P.W. *Molecules* (WH Freeman and Co, New York, 1987)

Ball, P. *Designing the Molecular World: Chemistry at the Frontier* (Princeton University Press, Princeton, NJ, 1994)

Borchardt-Ott, W. *Crystallography*, 4th edn (Springer, Berlin/ New York, 1993)

Bowser, J., *Inorganic Chemistry* (Brooks/Cole, Baltimore, 1993)

Brock, W.H. *The Fontana History of Chemistry* (Fontana, London, 1992)

Cowie, J.M.G., *Polymers: Chemistry and Physics of Modern Materials*, 2nd edn (Blackie Academic and Professional, Glasgow/New York, 1991)

Crick, F. *What Mad Pursuit* (Penguin, London 1990)

Donovan, A. *Antoine Lavoisier: Science, Administration and Revolution* (Blackwell, Oxford, UK/Cambridge, USA, 1994)

Elsom, D.M. *Atmospheric Pollution: A Global Problem*, 2nd edn (Blackwell, Oxford, UK/Cambridge, USA, 1992)

Emery, D. and Robinson, A. *Inorganic Geochemistry* (Blackwell Scientific, Oxford, UK/Cambridge, USA, 1994)

Emsley, J. *The Consumer's Good Chemicals Guide* (WH Freeman, Oxford/ New York, 1994

Frausto da Silva, J. and Williams, R. *The Biological Chemistry of the Elements* (Oxford University Press, Oxford/New York, 1993)

Gifford, C. *Essential Chemistry* (Usborne, London, 1992)

Gribben, J. *In Search of the Double Helix* (Black Swan, Ealing, 1993)

Hess, F.C. *Chemistry Made Simple* (W.H. Allen, London, 1971)

Hibbert, D. *Introduction to Electrochemistry* (Macmillan, New York, 1993)

Hill, G., Holman, J., Lazonby, J., Raffan, J., Waddington, D. *Chemistry: The Salters' Approach* (Heinemann Educational, Oxford/Portsmouth, NH, 1989)

Hornby, M. and Peach, J. *Foundations of Organic Chemistry* (Oxford University Press, Oxford/New York, 1993)

Jacques, J. *The Molecule and its Double* (McGraw Hill, New York, 1993)

Jaffe, B. *Crucibles: The Story of Chemistry From Ancient Alchemy to Nuclear Fission*, 4th edn (Dover, New York, 1976)

Mason, S. *Chemical Evolution* (Oxford University Press, Oxford/ New York, 1993)

O'Neill, P. *Environmental Chemistry*, 2nd edn (Chapman and Hall, London/New York, 1993)

Owen, S.M. and Brooker, A.T. *A Guide to Modern Inorganic Chemistry* (Longman, Harlow, Essex/John Wiley and Sons, New York, 1991)

Roberts, R.M. *Serendipity: Accidental Discoveries in Science* (Wiley, New York, 1989)

Scott, A. *Molecular Machinery: The Principles and Powers of Chemistry* (Blackwell, Oxford, UK/Cambridge, USA, 1989)

Selinger, B., *Chemistry in the Marketplace: A Consumer Guide*, 4th edn (Harcourt Brace Jovanovich, London, San Diego, 1989)

Skoog, D.A., West, D.M. and Holler, F.J. *Analytical Chemistry: An Introduction*, 6th edn (Saunders, Philadelphia/London, 1994)

von Baeyer, H.C. *Taming the Atom: The Emergence of the Visible Microworld* (Penguin, London and Random House, New York, 1992)

Warren, W.S. The *Physical Basis of Chemistry* (Academic Press, San Diego/ London, 1994)

Watson, J.D. *The Double Helix* (Penguin, London, 1970)

Winter, M.J. *Chemical Bonding* (Oxford University Press, Oxford/ New York, 1994)

The Nobel Prize for Chemistry, like similar prizes for Physics, Physiology or Medicine, Literature and Peace, has been awarded annually since 1901 as the world's most prestigious prize for outstanding work in the field, under the terms of the will of the Swedish chemist and engineer Alfred Nobel, who died in 1896. The prizes for physics and for chemistry are awarded by the Royal Swedish Academy of Sciences.

1901 **Jacobus Henricius van't Hoff** *Dutch*
Study of the laws of chemical dynamics and osmotic pressure

1902 **Emil Fischer** *German*
Organic synthesis, notably sugars and purine groups

1903 **Svante Arrhenius** *Swedish*
Development of the electrolytic theory of ionic dissociation

1904 **William Ramsay** *British*
Discovery of the rare elements helium, krypton, neon and xenon

1905 **Adolf von Baeyer** *German*
Development of organic dyes including indigo and study of hydroaromatic compounds

1906 **Henri Moissan** *French*
Discovery of fluorine and development of the electric furnace

1907 **Eduard Buchner** *German*
Research in biochemistry, including cell-free fermentation

1908 **Ernest Rutherford** *New Zealand*
Investigations into alpha particles, radioactive decay and the chemistry of radioactive substances

1909 **Wilhelm Ostwald** *German*
Study of chemical reactions and catalysts

1910 **Otto Wallach** *German*
Studies of alicyclic compounds

1911 **Marie Curie** *French*
Discovery of the elements radium and polonium

1912 **Victor Grignard and Paul Sabatier** *French*
Studies in organic synthetic chemistry

1913 **Alfred Werner** *Swiss*
Study of the linkage of atoms in molecules

1914 **Theodore Richards** *American*
Determination of the exact atomic weights of many elements

1915 **Richard Willstätter** *German*
Study of chlorophyll and its constituent pigments

1916–17 No award

1918 **Fritz Haber** *German*
Development of the process for the industrial synthesis of ammonia

1919 No award

1920 **Walther Nernst** *German*
Study of the heat of reactions (thermochemistry)

1921 **Frederick Soddy** *British*
Study of isotopes

1922 **Francis Aston** *British*
Mass spectroscopy and the discovery of isotopes of many non-radioactive elements

1923 **Fritz Pregl** *Austrian*
Development of microanalysis of organic substances

1924 No award

1925 **Richard Zsigmondy** *German*
Study of colloidal solutions

1926 **Theodore Svedberg** *Swedish*
Work on disperse systems

1927 **Heinrich Otto Wieland** *German*
Study of the constitution of the bile acids and related substances

1928 **Adolf Windaus** *German*
Study of steroids, notably sterols and their connection with vitamins

1929 **Arthur Harden** *British* **and Hans von Euler-Chelpin** *German-Swedish*
Study of enzymes and fermentation

1930 **Hans Fischer** *German*
Study of the structure of hemoglobin and chlorophyll

1931 **Carl Bosch and Friedrich Bergius** *German*
Ammonia synthesis and the hydrogenation of coal

1932 **Irving Langmuir** *American*
Investigations into surface chemistry

1933 No award

1934 **Harold Urey** *American*
Discovery of heavy hydrogen (deuterium)

1935 **Frédéric and Irène Joliot-Curie** *French*
Synthesis of new radioactive isotopes

1936 **Peter Debye** *Dutch*
Study of dipole moments, and the diffraction of X rays and electrons in gases

1937 **Walter Haworth** *British* **and Paul Karrer** *Swiss*
Investigations in vitamins

1938 **Richard Kühn** *German*
Research into vitamins and carotenoids. Forced to decline prize, which was finally awarded in 1946

1939 **Adolf Butenandt** *German* **and Leopold Ruzicka** *Swiss*
Studies of sex hormones and polymethylenes. Butendandt was forced to decline prize

1940-42 No award

1943 **György von Hevesy** *Hungarian-Swedish*
Use of isotopes as tracers in the study of chemical processes

1944 **Otto Hahn** *German*
Discoveries in nuclear fission of heavy elements

1945 **Artturi Virtanen** *Finnish*
Studies in agricultural biochemistry and food preservation

1946 **James Sumner, Wendell Stanley, and John Northrop** *American*
Preparation of pure enzymes and enzyme crystallization

1947 **Robert Robinson** *British*
Investigations in plant biochemistry, especially alkaloids

1948 **Arne Tiselius** *Swedish*
Study of serum proteins, electrophoresis and adsorption analysis

1949 **William Giauque** *American*
Cryogenic studies and chemical thermodynamics

1950 **Otto Diels and Kurt Adler** *West German*
Organic syntheses

1951 **Edwin McMillan and Glenn Seaborg** *American*
Discovery of the chemistry of transuranium elements

1952 **Archer Martin and Richard Synge** *British*
Invention of partition chromatography

1953 **Hermann Staudinger** *West German*
Discoveries in polymers, and theory of macromolecular chains

1954 **Linus Pauling** *American*
Research into the nature of the interatomic forces

1955 **Vincent du Vigneaud** *American*
Synthesis of a polypeptide hormone

△ Marie Curie (1911), who also won the prize for physics in 1903.

△ Hermann Staudinger (1953), pioneer of the chemistry of polymers and plastics.

1956 **Cyril Hinshelwood** *British* and **Nikolai Semenov** *Soviet*
Researches into the mechanisms of chemical chain reactions

1957 **Alexander Todd** *British*
Work on nucleotides and composition of cell proteins

1958 **Frederick Sanger** *British*
Work on the structure of insulin

1959 **Jaroslav Heyrovsky** *Czech*
Development of the analytic technique of polarography

1960 **Willard Libby** *American*
Development of radiocarbon dating technique

1961 **Melvin Calvin** *American*
Study of the processes of photosynthesis

1962 **John Kendrew and Max Perutz** *British*
Studies of the structure of globular proteins

1963 **Karl Ziegler** *West German* and **Giulio Natta** *Italian*
Study of polymers and polymerization reactions

1964 **Dorothy Hodgkin** *British*
X-ray analysis of large organic molecules

1965 **Robert Woodward** *American*
Synthesis of large organic compounds, including chlorophyll

1966 **Robert Mulliken** *American*
Work on the chemical bond and the electronic structure of molecules

1967 **Ronald Norrish, George Porter** *British*, and **Manfred Eigen** *West German*
Studies and measurement of extremely fast reactions

1968 **Lars Onsager** *American*
Discovery of non-equilibrium thermodynamics

1969 **Derek Barton** *British* and **Odd Hassel** *Norwegian*
Study of the effect of stereochemistry on reaction rates

1970 **Luis Leloir** *French-Argentinian*
Study of energy-storing biochemicals

1971 **Gerhard Herzberg** *Canadian*
Study of free radicals

1972 **Christian Anfinsen, Stanford Moore and William Stein** *American*
Contributions to enzyme chemistry

1973 **Ernst Fischer** *West German* and **Geoffrey Wilkinson** *British*
Study of the chemistry of organometallic sandwich compounds

1974 **Paul Flory** *American*
Development of analytic methods for the study of long-chain molecules

1975 **John Cornforth** *Australian-British* and **Vladimir Prelog** *Czech-Swiss*
Contributions to stereochemistry

1976 **William Lipscomb Jr** *American*
Study of the chemistry of boranes

1977 **Ilya Prigogine** *Belgian*
Contributions to non-equilibrium thermodynamics

1978 **Peter Mitchell** *British*
Study of energy transfer in cells

1979 **Herbert Brown** *American* and **Georg Wittig** *West German*
Preparation of organoboron compounds, useful in synthesis

1980 **Frederick Sanger** *British*, **Paul Berg and Walter Gilbert** *American*
Methods of determining the detailed structure and function of DNA

1981 **Kenichi Fukui** *Japanese* and **Roald Hoffmann** *American*
Application of quantum mechanics to predict the course of chemical reactions

1982 **Aaron Klug** *South African-British*
Development of crystallographic electron microscopy, and analysis of the structure of nucleic acid-protein complexes

1983 **Henry Taube** *Canadian-American*
Research into the transfer of electrons between metals in chemical reactions

1984 **Bruce Merrifield** *American*
Development of automated methods of assembling peptides to synthesize proteins

1985 **Herbert Hauptman and Jerome Karle** *American*
Development of a rapid method of determining the structures of biochemical molecules from X-ray diffraction patterns

1986 **Dudley Herschbach** *American*, **Yuan Lee** *Chinese-American* and **John Polanyi** *Canadian*
Work on reaction dynamics

1987 **Donald Cram** *American*, **Charles Pedersen** *Norwegian-American* and **Jean-Marie Lehn** *French*
Work on synthetic molecules that can mimic chemical reactions of life processes

1988 **Johann Diesenhofer, Robert Huber and Hartmut Michel** *West German*
Research into photosynthesis

1989 **Sidney Altman and Thomas Cech** *American*
Discovery of the catalytic action of RNA in cellular reactions

1990 **Elias Corey** *American*
Development of techniques for artificially duplicating natural substances for use as drugs

1991 **Richard Ernst** *Swiss*
Improvements in the use of nuclear magnetic resonance (NMR) to analyze chemicals

1992 **Rudolph Marcus** *Canadian*
Prediction of the interactions between molecules in a solution

1993 **Kary Mullis** *American* and **Michael Smith** *British*
Work on polymerase chain reactions and mutations in nucleotides

1994 **George Olah** *Hungarian-American*
Studies of super acid-carbocation chemistry

△ Dorothy Hodgkin (1964), discoverer of the structures of insulin and penicillin.

Acknowledgments

Picture credits

1 SPL/Jim Amos **2** AOL/OCD **3** Anthony Blake Picture Library, Richmond **4** AOL **6** IP/Martin Black **7** RHPL/IPC Magazines/Martyn Thompson **48–49** SPL/Philippe Plailly **50–51** SPL/Philippe Plailly **50b** TRH Pictures, London **51br** Z **51bl** PEP/Peter Scoones **54–55** SPL/Dr Jeremy Burgess **55** SPL/Sinclair Stammers **56bl** Paul Brierley, Hertfordshire **56br** Dr Beers Consolidated Mines **56–57** SPL/Geoff Tompkinson **58** inset HL **58–59** Z **60–61** Z/Runk/Schoenberger **62–63** SPL/Jeremy Burgess **63t** TCL/Susan Griggs **63b** TCL/Susan Griggs **64bl** SPL/Alex Bartel **64br** SPL/Alex Bartel **64–65** Anthony Blake Picture Library, Richmond **65bl** SPL/Alex Bartel **65br** SPL/Alex Bartel **66–67t** Still Pictures/Jorgen Schytte **66–67b** AOL **68–69** RF **69** Associated Press, London **70–71** Spectrum Colour Library/David Bailey **71** EPL/Matt Sampson **72** SPL/World View/Jaap Bouman **72–73** SPL/James Holmes, Hays Chemicals **73** RHPL/Walter Rawlings **74–75** SPL/Martin Bond **75t** Andree Abecais **75c** Nature Photographers, Hampshire **76–77** Galvanisers Association **79** SPL/ Jim Amos **80–81** Z/Stockmarket **82–83** SPL/Martin Bond **84–85t** SPL/Dr Jeremy Burgess **84–85b** PEP/Peter Scoones **85** IP/ Rolf Hayo **86** Z **86–87** Spectrum Colour Library **87** AOL/OCD **88** IP/Rupert Conant **88–89** AOL **89** HL/Tony Hardwell **90–91b** RF/Peter Brooker **90–91t** AOL/OCD **92** AOL **92–93** IP/ Martin Black **94–95** Colorsport **96–97** FSP/Liaison/Giboux **97** AOL **98–99** Colorsport/Lacombe **101** Network Photographers/ Barry Lewis **102–103** Z/Stockmarket **103** RF/John Gooch **104** Allsport/David Cannon **104–105t** Z **104–105b** AOL **105** Colorsport **106** SPL/Peter Ryan **106–107** SPL/Alex Bartel **107** SPL/ Hank Morgan **108–109** SPL/J C Revy **110–111** SPL/ Oxford Molecular Biophysics Laboratory **112–113** SPL/J C Revy **113** Ardea/Adrian Warren **114–115** BCL/M P Kahl **115bl** SPL/ CNRI **115br** Colorific/John Moss **116–117** Z **117t** SPL/James Prince **117b** HL/Robert Francis **118** RF/Today **118–119** Colorsport/William R Sallaz **119** Colorsport/Bardon Sport **120–121** Z **122** AOL **122–123** ACE Photo Agency/Auschromes **123bl** AOL **123tr** AOL **123br** AOL **125** IP/David Palmer **126** RHPL/IPC Magazines/Martyn Thompson **126–127t** TIB/Paul Trummer **126–127b** AOL **128l** TIB/David de Lossy **128b** Science & Society Picture Library **128–129** Christopher Joyce **130–131** NHPA/David Woodfall **134–135** Z/Kelly/Mooney **138–139** SPL/Geoff Tompkinson **140–141** AOL **154** SPL/National Library of Medicine **155t** Popperfoto **155b** Hulton Deutsch Collection

Abbreviations

b = bottom, **t** = top
l = left, **c** = center, **r** = right

AOL	Andromeda Oxford Limited, Abingdon, UK
EPL	Environmental Picture Library, London, UK
HL	Hutchison Library, London, UK
IP	Impact Photos, London, UK
NHPA	Natural History Photographic Agency, Sussex, UK
PEP	Planet Earth Pictures, London, UK
RF	Rex Features, London, UK
RHPL	Robert Harding Picture Library, London, UK
SPL	Science Photo Library, London, UK
TCL	Telegraph Colour Library, London, UK
Z	Zefa Picture Library, London, UK

Artists

Mike Badrock, Rob and Rhoda Burns, John Davies, Hugh Dixon, Bill Donohoe, Sandra Doyle, John Francis, Shami Ghale, Mick Gillah, Ron Hayward, Jim Hayward, Trevor Hill/Vennor Art, Joshua Associates, Frank Kennard, Pavel Kostell, Ruth Lindsey, Mike Lister, Jim Robins, Colin Rose, Colin Salmon, Leslie D. Smith, Ed Stewart, Tony Townsend, Halli Verinder, Peter Visscher

Editorial assistance

Peter Lafferty, Ray Loughlin, Trevor Pryce-Jones, Lin Thomas

Index

Ann Barrett

Origination by

HBM Print Ltd, Singapore;
ASA Litho, UK